Fire Detection and Suppression Systems

Third Edition

Lynne Murnane
Project Manager/Editor

Ted Boothroyd
Technical Writer

Tom Ruane
Senior Editor

Validated by the International Fire
Service Training Association
Published by Fire Protection Publications
Oklahoma State University

RECYCLABLE

The International Fire Service Training Association

The International Fire Service Training Association (IFSTA) was established in 1934 as a "nonprofit educational association of fire fighting personnel who are dedicated to upgrading fire fighting techniques and safety through training." To carry out the mission of IFSTA, Fire Protection Publications was established as an entity of Oklahoma State University. Fire Protection Publications' primary function is to publish and disseminate training texts as proposed and validated by IFSTA. As a secondary function, Fire Protection Publications researches, acquires, produces, and markets high-quality learning and teaching aids as consistent with IFSTA's mission.

The IFSTA Validation Conference is held the second full week in July. Committees of technical experts meet and work at the conference addressing the current standards of the National Fire Protection Association and other standard-making groups as applicable. The Validation Conference brings together individuals from several related and allied fields, such as:

- Key fire department executives and training officers
- Educators from colleges and universities
- Representatives from governmental agencies
- Delegates of firefighter associations and industrial organizations

Committee members are not paid nor are they reimbursed for their expenses by IFSTA or Fire Protection Publications. They participate because of commitment to the fire service and its future through training. Being on a committee is prestigious in the fire service community, and committee members are acknowledged leaders in their fields. This unique feature provides a close relationship between the International Fire Service Training Association and fire protection agencies which helps to correlate the efforts of all concerned.

IFSTA manuals are now the official teaching texts of most of the states and provinces of North America. Additionally, numerous U.S. and Canadian government agencies as well as other English-speaking countries have officially accepted the IFSTA manuals.

ISBN 0-87939-267-3   Library of Congress Control Number: 2005933573

Third Edition, First Printing, December 2005   *Printed in the United States of America*

10   9   8   7   6   5   4   3

If you need additional information concerning the International Fire Service Training Association (IFSTA) or Fire Protection Publications, contact:

Customer Service, Fire Protection Publications, Oklahoma State University
930 North Willis, Stillwater, OK 74078-8045
800-654-4055   Fax: 405-744-8204

For assistance with training materials, to recommend material for inclusion in an IFSTA manual, or to ask questions or comment on manual content, contact:

Editorial Department, Fire Protection Publications, Oklahoma State University
930 North Willis, Stillwater, OK 74078-8045
405-744-4111   Fax: 405-744-4112   E-mail: editors@osufpp.org

Table of Contents

Preface.. vii

Introduction... 1

Purpose and Scope 3

Manual Organization 3

Key Information.................................... 4

1   Fire Extinguishing Agents and Portable Fire Extinguishers 7

Extinguisher Symbols 7

Pictorial System..8

Letter-Symbol System 9

How Extinguishers are Rated.......... 9

General Rating Criteria............................... 11

Class A Rating Tests.................................... 12

Class B Rating Tests.................................... 12

Class C Rating Tests.................................... 12

Class D Rating Tests.................................... 12

Class K Rating Tests.................................... 14

Extinguishing Agents 15

Water.. 15

Antifreeze.. 15

Alkaline Mixtures....................................... 15

Carbon Dioxide... 16

Aqueous Film Forming Foam (AFFF)......... 16

Film Forming Fluoroprotein (FFFP)........... 17

Halons.. 18

Halon Replacement Agents 18

Dry Chemical Agents 19

Types of Fire Extinguishers............ 23

Stored-Pressure Extinguishers 23

Cartridge-Operated Extinguishers............. 23

Pump-Operated Extinguishers.................... 24

Selection and Distribution of Extinguishers 24

Nature-of-the Hazard Factor............................. 26

Size-of-the-Extinguisher Factor 26

Class A Extinguisher Distribution Factors 27

Class B Extinguisher Distribution Factors 28

Class C and Class D Extinguisher Distribution Factors..................................29

Class K Extinguisher Distribution Factors 29

Installation and Placement of Extinguishers 29

Portable Fire Extinguishers on Fire Apparatus 31

Inspecting, Maintaining, and Recharging Extinguishers 31

Inspecting Extinguishers 32

Maintaining Extinguishers 34

Recharging Extinguishers 35

Hydrostatic Testing of Portable Extinguishers 36

Using Portable Extinguishers 39

General Techniques 39

P.A.S.S. Method... 42

Attacking Class A Fires............................... 43

Attacking Class B Fires............................... 45

Attacking Class C Fires 46

Attacking Class D Fires 46

Attacking Class K Fires............................... 47

Summary..47

2   Fire Detection and Signaling Systems 51

Basic System Components 51

Fire Alarm Control Unit.............................. 52

Power Supply... 52

Initiating Devices 54

Notification Appliances 55

Auxiliary Services..55

Types of Signaling Systems 56

Local System Protected Premises (Local) 57

Auxiliary Fire Alarm System 61

Remote Receiving System........................... 61

Proprietary System......................................62

Central Station System 62

Emergency Voice/Alarm
Communications System 63

Manual Alarm-Initiating Devices.................. 64

General Requirements...............................64

Coded Versus Noncoded Pull Stations.......65

Single-Action Pull Station 65

Double-Action Pull Station 65

Automatic Alarm-Initiating Devices 66

Fixed-Temperature Heat Detector.............66

Rate-of-Rise Heat Detector.......................67

Smoke Detector 67

Flame Detector 74

Fire Gas Detector 75

Combination Detector 76

**Inspecting and Testing Fire Detection and
Signaling Systems 76**

Acceptance Testing 76

System Service Testing and Periodic
Inspections 78

Record Keeping 81

Written Records 82

Electronic Records 83

Summary... 84

3   Introduction to Water Supply................. 87

Characteristics of Water 87

Extinguishing Properties of Water 87

Advantages and Disadvantages of Water......... 92

Understanding Water Pressure..................... 92

Principles of Pressure 94

Types of Pressure......................................96

Pressure Loss and Gain: Elevation and
Altitude..99

Friction Loss.. 99

Principles of Friction Loss 100

Reducing Friction Loss 103

Water Hammer 103

**Principles of Municipal Water Supply
Systems 104**

Sources of Water Supply 104

Means of Moving Water 104

Processing or Treatment Facilities 106

Water Distribution System 107

Private Water Supply Systems 115

**Water Supply Requirements for Standpipe
and Hose Systems 117**

**Water Supply Requirements for Automatic
Sprinkler Systems 118**

Pipe Schedule Systems............................118

Hydraulically Designed Water-Based
Extinguishing Systems 118

Duration of Water Supplies 118

Summary...119

4   Fire Pumps 123

Common Fire Pump Types 123

Horizontal Split-Case Pumps 123

Vertically Mounted Split-Case Pumps 125

Vertical-Shaft Turbine Pumps 125

Pump Drivers 127

Electric Motors 127

Diesel Engine Drivers.............................129

Steam Turbines 131

Pump Controllers 131

Controllers for Electric Motor-Driven
Pumps..131

Diesel Engine Controllers........................133

Pump Components and Accessories.............. 134

Pipe and Fittings.....................................134

Relief Valves 136

Circulation Relief Valve 136

Test Equipment.......................................136

Pressure Maintenance Pumps..................137

Gauges 138

Component Arrangement........................138

Standard Performance Specifications 139

Testing, Inspection, and Maintenance of Fire Pumps **139**

    Testing Fire Pumps.................................. 140

    Routine Operation and Maintenance 147

Summary **148**

5   Standpipes and Hose Systems **151**

Classification of Standpipe Systems **151**

    Class I Standpipe Systems 152

    Class II Standpipe Systems....................... 152

    Class III Standpipe Systems 152

Types of Standpipe Systems **153**

Fire Department Connections **153**

Water Supply Considerations **155**

    Standpipes... 155

    Hoselines 156

    Sprinkler Systems 156

Water Pressure Considerations **157**

    Pressure-Regulating Devices 158

Inspecting and Testing Standpipes.................**159**

    Initial Installation Inspection and Tests 159

    In-Service Inspections............................. 159

Summary..**160**

6   Automatic Sprinkler Systems............... **165**

Components of Sprinkler Systems...................**166**

    Sprinklers 168

    Temperature Ratings................................. 168

    Sprinkler Response Time........................... 169

    Deflector Component 171

    Early-Suppression Fast-Response Sprinklers.. 172

Sprinkler System Piping..................................**173**

Valves..**175**

    Control Valves... 175

    Operating Valves 177

Fire Department Connections (FDCs).............**179**

Types of Sprinkler Systems **181**

Wet-Pipe Sprinkler System 181

Dry-Pipe Sprinkler System 185

Deluge Sprinkler System 189

Preaction Sprinkler System..................... 191

Sprinkler Systems for Storage Occupancies 192

Inspecting and Testing Sprinkler Systems **194**

    Sprinklers... 195

    Sprinkler Piping, Hangers, and Seismic Braces 196

    Changes in Building Occupancy 197

    Inspecting and Testing Wet-Pipe Sprinkler Systems................................. 198

    Inspecting and Testing Dry-Pipe Sprinkler Systems.................................203

    Inspecting and Testing Deluge and Preaction Sprinkler Systems.................208

Restoring Sprinkler Systems............................**210**

    Restoring Wet-Pipe Sprinkler Systems..... 210

    Restoring Dry-Pipe Sprinkler Systems..... 210

    Restoring Deluge and Preaction Sprinkler Systems................................. 210

Sprinkler System Impairment Control **210**

Residential Sprinkler Systems **211**

    Residential Sprinklers Versus Conventional Sprinklers 212

    Residential Sprinkler Piping 212

    Water Supply and Flow Rate Requirements 213

    Residential Sprinkler Spacing................... 214

Summary..**215**

7   Special Extinguishing Systems........... **219**

Wet Chemical Extinguishing Systems **219**

    System Design..220

    Inspection and Test Procedures 221

Dry Chemical Extinguishing Systems **222**

Gaseous Systems **223**

    Carbon Dioxide Extinguishing Systems..223

Foam Extinguishing Systems........................ **226**

  System Description 227

  How Foam Extinguishes Fire 230

  How Foam Is Generated 230

  Foam Proportioners 231

  Foam Nozzles And Sprinklers 236

  Foam System Inspection And Testing 241

Summary.. **241**

Appendix A NFPA Job Performance
  Requirements with Page References 245

Appendix B Extinguisher Rating Tests 247

Appendix C Pump Tests 253

Appendix D Fire Department Operations
  with Standpipes 257

Appendix E Fire Department Operations
  at Sprinklered Occupancies 263

Appendix F Design Considerations for
  Automatic Sprinkler Systems 269

Appendix G Storage Methods Used in
  Warehouses 273

Glossary 279

Index 285

Preface

This 3rd edition of **Private Fire Protection and Detection** has been retitled to better reflect its contents. **Fire Detection and Suppression Systems** is intended to familiarize fire department and industrial fire protection personnel with the various types of fire protection systems found in public and private buildings.

Acknowledgment and special thanks are extended to the members of the material review committee who contributed their time, wisdom, and knowledge to the development of this manual.

IFSTA Fire Detection and Suppression Systems 3rd Edition Validation Committee

Committee Chair
Bradd Clark, Fire Chief
Claremore (OK) Fire Department

Paul Boecker III, Firefighter/EMT
Sugar Grove (IL) Fire District

Rick McIntyre, Fire Chief
Jacksonville (FL) Fire Department

Kristofer DeMauro, Fire Marshal
City of Owasso (OK) Fire Department
8901 North Garnett

Andy Miller, Asset Protection Administrator
Hallmark Cards
Kansas City, MO 64141-6580

Ed Hawthorne, Manager
Shell Oil Company
Deer Park, TX 77536

Edward Prendergast
Chicago, IL 60643

J.D. Rice, Fire Chief
Valdosta (GA) Fire Department

Brett Lacey, Senior Fire Protection Engineer
Colorado Springs (CO) Fire Department

The work of Technical Writer Ted Boothroyd is much appreciated. He thoroughly and patiently researched and updated the entire draft. The committee's work was much easier as a result of his efforts.

Grateful thanks is also extended to the following organizations and individuals who contributed information, photographs, and technical assistance that were instrumental in the development of this manual:

Chris E. Mickal, New Orleans Fire Department Photo Unit, LA

Professor Pat D. Brock and students of the Fire Protection and Safety Technology Program, OSU

Dr. Tom Woodford, Associate Professor and Department Head, School of Fire Protection and Safety Technology, OSU

Floyd Luinstra, Adjunct Professor, School of Fire Protection and Safety Technology, OSU

Cody Spybuck, Environmental Health and Safety Services, OSU

Roy Mason, Environmental Health and Safety Services, OSU

Randy Kelley, Environmental Health and Safety Services, OSU

Paul Ramirez, Phoenix (AZ) Fire Department

Trent Hawkins, Fire Marshal, Stillwater (OK) Fire Department

Jim Feld, Fire Protection Engineer, Fairfield, CA

Santa Rosa Fire Equipment, Santa Rosa, CA

Fire Service Training, Oklahoma State University

Jason Louthan	Dan Knott
Richard Teeter	Brian West
Jerimiah Hoffstatter	

Bart Foster, MerCruiser Sterndrives and Inboards, Stillwater, OK

Tyco Safety Products

Amerex Corporation

FlexHead Industries

MerCruiser, Stillwater, Oklahoma

Chemetron Fire Systems

Last, but certainly not least, gratitude is extended to the members of the Fire Protection Publications staff who contributed their time and talent to the completion of this manual.

Jeff Fortney, Senior Editor

Cynthia Brakhage, Senior Editor

Tara Gladden, Editorial Assistant

Don Davis, Production Manager

Ann Moffat, Production Coordinator

Clint Parker, Senior Graphic Designer

Ben Brock, Senior Graphic Designer

Jim Austell, Warehouse Foreman

David Caldwell, Stock Clerk III

Jeff Blocker, Stock Clerk III

Scott Walker, Warehouse Foreman

Kelly Pickering, Warehouse Manager

Introduction

Fire detection and suppression systems are installed in a multitude of occupancies with an equal amount of personnel interested or responsible for them. The purpose of fire detection and suppression systems is to provide a means of notification to occupants or fire personnel and to initiate the earliest attempts to control or suppress the fire. These devices and equipment can be either manually operated or automatically operated. A key aspect of such devices and equipment is having *trained* personnel to rely on and to take appropriate action in the event of a fire emergency. (**NOTE:** More information on the preparation of employees to handle fire protection duties is contained in IFSTA's **Industrial Emergency Services Training: Incipient Level** manual.) However, no single manual can address all of the many varieties of protective devices, systems, or equipment that may be encountered in the business and industrial community. This manual is intended to be used as a teaching tool to be used by fire service personnel, instructors, and individuals employed in private industry. The manual reflects the recommendations of many agencies recognized by the vast majority of fire protection professionals to be the minimum standard. These agencies include the National Fire Protection Association (NFPA), Underwriters Laboratories, FM Global, and other credible resources.

NOTE: This manual on fire detection and suppression systems is not intended to substitute for any codes, standards, or policies and should not be used as the authority for examination of any legally constituted requirement.

The most common fire protection device is the portable fire extinguisher. Portable fire extinguishers are widely considered to be the most basic and first line defense tool against an incipient fire emergency. Fire prevention and suppression personnel, as well as maintenance personnel, should have an intimate knowledge of the characteristics of the fuel burning and the application of the proper extinguishing agent. This manual will cover the selection of the appropriate portable fire extinguisher for a given hazard, including the new Class K designation for commercial cooking equipment.

Fire detection and signaling systems are used to alert building occupants and/or organized fire protection units, and are also used to activate fire protection system components. Detection systems use some type of device that is sensitive to one or more products of combustion. Automatic fire detection and signaling systems may be required by codes and ordinances to be installed in certain buildings for protection of life. These systems are especially important in isolated and/or high life hazard facilities where automatic sprinkler protection is not provided. With more emphasis being placed on early detection and standards requiring detectors to be installed in dwellings, it is becoming increasingly necessary for all fire service professionals to become well-versed in these devices and the various products of combustion that activate them.

Water Supply is a new chapter added to this manual. It is written with fire protection systems in mind, not water supply as it relates to fire apparatus. Both automatic sprinkler systems and standpipe systems are dependent on adequate supplies of water to be effective. An adequate supply of water is one that is sufficient in both volume and pressure. In cases where the building is large or tall, the municipal water supply may not be adequate in itself to meet the demands of the system. In these cases, it is necessary to provide on-site water supply sources and/or fire pumps to feed the system. Chapter 3 discusses the principles of water pressure, water distribution, and water flow analysis. Its content may be useful to fire suppression and prevention personnel, fire service instructors, and building representatives or owners.

Industrial fire pumps and their various applications are discussed in Chapter 4, addressing the three primary types of pumps and their use as either booster pumps or fire pumps. The chapter further covers pump drivers (the motor or engine), pump controllers (the control panel), and the many components and accessories commonly found in the field. Because industrial fire pumps may not be used in live situations very often, these pumps should be tested regularly using a set procedure. A fire inspector, college instructor, or building official will find useful information on the subject of testing, inspection, and maintenance of industrial fire pumps as well as water supply requirements. References are made to other IFSTA publications or NFPA standards when applicable.

Although they cannot take the place of automatic sprinkler systems, standpipe and hose systems are designed to provide a quick and convenient means for operating fire streams on all levels of building construction. Depending on the type installed, the standpipe system may be used by firefighters, by building occupants, or by both. Standpipe systems should be ready for use at all times. The chapter on standpipe systems will prepare the fire professional to recognize the state of readiness of the system, as well as provide a basic understanding of the piping and fixtures involved.

Considered by many to be the premium line of defense against fires, automatic sprinkler systems in their basic form have been used for over 100 years. Their origins date back to the days of the large industrial mills in the northeastern United States during the Industrial Revolution. Today, the automatic sprinkler system remains an unsurpassed fire protection device. Annual records reveal that in buildings where automatic sprinklers were installed, a large percent of all fires were controlled or extinguished by these systems. Automatic sprinkler systems can be designed and tailored to the specific needs required of a particular occupancy. Chapter 6 of this manual discusses all common automatic sprinkler systems as well as recent innovations and recommendations to the systems.

Special-agent fixed-extinguishing systems are used in those situations where automatic water sprinkler systems are not desirable. In these instances, protection must still be provided safely and effectively. This can be accomplished by the use of carbon dioxide (CO_2), halogenated agents or Halon substitutes, or dry-chemical fixed systems. Certain hazards or compromises are imposed when using agents other than water. The type of system used, as well as its arrangement, will depend on such factors as the hazard, its size, its organization, and economics. This manual covers the applications of the most common special extinguishing agents and the advantages of their use.

A vast number of variations in types, makes, and models of fire protection and detection equipment are recognized. Systems, equipment, and procedures described herein have been used to illustrate typical installations. Generally, the more complicated or sophisticated a system, the more important skillful maintenance becomes. It is recommended that the manufacturer's technical data be consulted for specific answers to questions pertaining to design, installation, operation, and maintenance of components or systems.

The installation, maintenance, and testing of fire protection and detection equipment must be done in compliance with local, state, and federal codes, ordinances, and standards. Many installations also follow applicable guidelines by agencies discussed earlier.

Fire service professionals will find that materials in this manual will help to bring about a better understanding of fire detection and suppression system installations. Those personnel involved with fire protection management, safety, and maintenance should also be better able to understand the concern of the public fire services for proper maintenance of this equipment.

Often, building codes and ordinances are relaxed due to the installation of fire protection and detection systems. This could constitute a serious threat to the safety of public and fire service personnel in the event a fire occurs.

Purpose and Scope

This third edition of **Fire Detection and Suppression Systems** (earlier editions were titled **Private Fire Protection and Detection**) is designed to provide up-to-date information on fire detection and suppression systems. It is aimed at municipal firefighters, industrial fire protection personnel, and all others seeking additional information in this area. It contains information on automatic sprinkler systems, hose standpipe systems, and fixed fire pump installations. Portable fire extinguishers, fixed special agent extinguishing systems, and fire alarm and detection systems are addressed. Information on the design, operation, maintenance, and inspection of these systems and equipment is provided.

> This manual is intended to familiarize the fire service and other interested personnel with hazard recognition, extinguishing agents, detection and alarm systems, fixed extinguishing systems, portable extinguishers, fire behavior, and water supplies for extinguishing systems. This manual is not intended to substitute for training as a fire inspector or for systems maintenance and inspection training.

Manual Organization

This manual is divided into seven chapters:

- Portable Extinguishers
- Detection and Signaling Systems
- Introduction to Water Supply
- Fire Pumps
- Standpipe and Hose Systems
- Sprinkler Systems
- Special Extinguishing Systems

Also included is a glossary at the end of the text and appendices that are designed to help the reader further understand testing, fire fighting, or administrative procedures related to fire detection equipment and systems.

NOTE: For more information on training personnel in using these systems and on providing fire protection services to private facilities, see the following manuals:

- IFSTA **Industrial Emergency Services Training: Incipient Level**
- IFSTA **Fire Inspection and Code Enforcement**
- FPP *Fire Protection Hydraulics and Water Supply Analysis*

Learning Objectives
IFSTA Learning Objectives

Learning objectives have been developed for this manual. They are found at the beginning of each chapter and may be used by the readers to guide their study of the subjects addressed in those chapters.

NFPA Standards

This manual addresses certain objectives related to these topics contained in the following National Fire Protection Association standards:

- NFPA 1001, *Standard for Fire Fighter Professional Qualifications* (2002)
- NFPA 1002, *Standard for Fire Apparatus Driver/ Operator Professional Qualifications,*
- NFPA 1031, *Standard for Professional Qualifications for Fire Inspector and Plan Examiner,* 2003 Edition.
- NFPA 10, *Standard for Portable Fire Extinguishers,* 2002 Edition.
- NFPA 13, *Standard for the Installation of Sprinkler Systems,* 2002 Edition.
- NFPA 20, *Standard for the Installation of Stationary Pumps for Fire Protection,* 2003 Edition.
- NFPA 22, *Standard for Water Tanks for Private Fire Protection,* 2003 Edition.
- NFPA 24, *Standard for the Installation of Private Fire Service Mains and Their Appurtenances,* 2002 Edition.
- NFPA 25, *Standard for the Inspection, Testing, and Maintenance of Water-Based Fire Protection Systems,* 2002 Edition.

NOTE: Many of the IFSTA learning objectives address knowledge or skills outlined in NFPA standards. See **Appendix A,** NFPA Job Performance Requirements (JPRs) with Chapter References, for a listing of JPRs addressed in each chapter.

FESHE Objectives

Portions of this manual address the objectives found in the National Fire Academy's Fire and Emergency Services Higher Education (FESHE) Fire Science Curriculum, *Fire Protection Systems* course. References to these objectives can be found at the beginning of the chapter in which they are discussed.

Key Information

Various types of information in this book are given in shaded boxes marked by symbols or icons. See the following examples:

Information

Information boxes give facts that are complete in themselves but belong with the text discussion. It is information that may need more emphasis or separation. They can be summaries of points, examples, calculations, scenarios, or lists of advantages/disadvantages.

Information Plus

Additional relevant information that is more detailed, descriptive, or explanatory than that given in the text.

Safety Reminder

Contains information that is designed to accent information found in the general text.

Three key signal words are found in the text: WARNING, CAUTION, and NOTE. Definitions and examples of each are as follows:

- **WARNING** indicates information that could result in death or serious injury to fire and emergency services personnel. See the following example:

> ## WARNING!
> Do not confuse an ordinary water extinguisher with a "water mist" extinguisher. A water mist extinguisher is rated for use on a Class C fire but a plain water fire extinguisher is not. Using the wrong type of extinguisher can result in severe harm.

- **CAUTION** indicates important information or data that fire and emergency service responders need to be aware of in order to perform their duties safely. See the following example:

> ## CAUTION!
> Foam extinguishers are only useful on relatively small flammable liquid fires. Foam does not provide much radiant heat protection, so the extinguisher operator is at serious risk of injury if the fire is too big to be handled by the extinguisher. A dry chemical extinguisher would be a much better choice in that situation.

- **NOTE** indicates important operational information that helps explain why a particular recommendation is given or describes optional methods for certain procedures. See the following example:

NOTE: The material in this chapter is intended to be informative and descriptive, but should not be used as the authority over local codes, ordinances, and standards.

Fire Extinguishing Agents & Portable Fire Extinguishers

1. Describe the different symbol systems used on portable extinguishers.

2. Identify five classifications of portable fire extinguishers.

6. Identify the types of extinguishing agents used in portable fire extinguishers.

3. Describe the methods by which extinguishers are rated.

4. Discuss the advantages and disadvantages of different extinguishing agents.

5. Describe the operating principles of different types of fire extinguishers.

6. Identify life safety hazards associated with certain extinguishing agents.

7. Describe the method of determining the proper distribution of portable fire extinguishers.

8. Identify the basic elements of a portable fire extinguisher maintenance examination.

9. Discuss general procedures for inspecting and testing fire extinguishers.

10. Describe the general techniques for the use of portable fire extinguishers on different classes of fires.

FESHE Objectives

Fire and Emergency Services Higher Education (FESHE) Objectives:
Fire Science Curriculum: Fire Protection Systems

•   Identify and describe appropriate national standards governing the installation, inspection, and maintenance of given extinguishing agent/systems and their related components.

Chapter 1
Fire Extinguishing Agents and Portable Fire Extinguishers

A portable fire extinguisher is viewed by some as the first line of defense against incipient fires of limited size. A fire extinguisher should never be viewed as a substitute for automatic fire detection and suppression systems but rather as a complement to them. The Occupational Safety and Health Administration (OSHA) of the United States Department of Labor, in cooperation with the National Association of Fire Equipment Distributors (NAFED), conducted a study in 1996 that was designed to determine the effectiveness of fire extinguishers in certain occupancies. The study provided statistical information that supported the fact that a portable extinguisher enables an individual with minimal training and orientation to extinguish an incipient fire. Of the 2,267 fires reported during the study, portable fire extinguishers successfully extinguished 2,161 fires (95.3 percent). In 2,088 of these reported fires (92.1 percent), a single fire extinguisher was sufficient to extinguish the fire.

The value of a fire extinguisher lies in the speed with which personnel not trained as firefighters can activate it and use it. For a portable fire extinguisher to be effective, the following requirements must be met:

- The extinguisher must be readily accessible.

- The extinguisher must be in working order.

- The extinguisher must be suitable for the hazard being protected.

- The person using the extinguisher must know how to operate it.

NOTE: Consult NFPA 10, *Standard for Portable Fire Extinguishers,* for more information on selecting and operating fire extinguishers.

This chapter provides information on the basic components of fire extinguishers and discusses the ways in which they are rated, tested, used, and inspected. Also covered are the agents used in the extinguishers and extinguisher placement with respect to fire hazards and fire codes.

Extinguisher Symbols

Fires have been broadly grouped into five classifications according to the burning characteristics of various combustible materials **(Figures 1.1 a-e, p. 8)**:

- Class A Fire – Fire involving ordinary combustibles such as wood, paper, and cloth

- Class B Fire – Fire involving flammable and combustible liquids and gases such as gasoline, kerosene, and liquefied petroleum gases

- Class C Fire – Fire involving energized electrical equipment

- Class D Fire – Fire involving combustible metals such as magnesium, sodium, and titanium

- Class K Fire – Fires involving cooking devices that contain or use combustible vegetable or animal oils and fats

NOTE: This most recent classification of fire was added by NFPA to reflect the development of cooking oils that require special extinguishing materials. In Europe this classification of fire is designated Class E.

Because the burning characteristics of various materials depend upon their physical and chemical properties, fire extinguishers are filled with a variety of extinguishing agents. It is critical that

Figure 1.1 Each type of fire hazard has different extinguisher requirements. Class A fires are effectively controlled with water. For fires involving Class B fuels, excluding the oxygen is the most effective extinguishing method. Class C fires, which involve energized electrical equipment, can be extinguished more rapidly once the equipment is de-energized. Fires involving Class D (combustible metal) materials may require specialized equipment. Class K fires involve cooking oils that are difficult to cool.

the operator be aware of the type of fire and the extinguisher type needed to combat it in order to be able to select the proper extinguisher. To simplify the process of matching different types of extinguishers with types of fires, a method for identifying extinguishers through the use of symbols has been developed. NFPA 10, *Standard For Portable Fire Extinguishers,* recognizes two methods of extinguisher recognition: the pictorial system and the letter-symbol system.

Pictorial System

The international "picture symbol" labeling system, which was designed by the National Association of Fire Equipment Distributors (NAFED), is the

most widely used fire extinguisher identification system. The system is designed to make the selection of fire extinguishers easier through the use of picture symbols. The symbols indicate the type of fire the extinguisher is capable of extinguishing. They also indicate when *not* to use an extinguisher on certain types of fires.

If an extinguisher is suitable for use on a particular class of fire, the picture-symbol background is light blue or black. If the extinguisher is not suitable for a particular class of fire, the picture symbol has a black background with a diagonal red line through the extinguisher symbol (**Figure 1.2**). Extinguishers are often suitable for more than one class of fire.

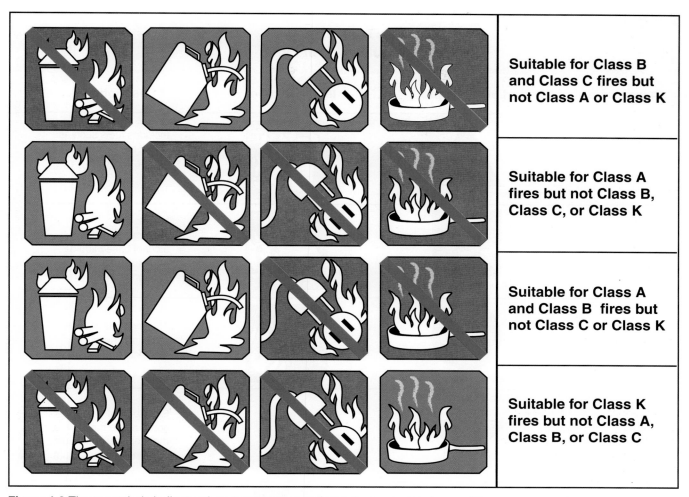

	Suitable for Class B and Class C fires but not Class A or Class K
	Suitable for Class A fires but not Class B, Class C, or Class K
	Suitable for Class A and Class B fires but not Class C or Class K
	Suitable for Class K fires but not Class A, Class B, or Class C

Figure 1.2 These symbols indicate when *not* to use an extinguisher on certain types of fires.

NOTE: This pictorial system applies only to Class A, Class B, Class C, and Class K fire. There is no pictorial symbol for Class D, combustible metal fires.

Letter-Symbol System

The letter-symbol method of extinguisher identification is older than the picture-symbol method. In the letter-symbol method, each class of fire is represented by its appropriate letter. The letters A, B, C, D, or K are enclosed by a particular geometric shape. In addition, the background of the geometric shape can be color-coded to further identify the extinguisher. It is important to note that the letter symbol method of fire extinguisher identification is rapidly losing its usefulness in favor of the internationally recognized pictorial system. The most recent symbol being used for Class K is the letter K within a hexagonal geometric shape. The

hexagonal shape, however, is not universal among fire extinguisher manufacturers and is likely to disappear along with other letter symbols in the future. See **Table 1.1, p. 10,** for a depiction of the letter-symbol identification method.

How Extinguishers are Rated

A portable fire extinguisher is rated according to its intended use and fire fighting capability on the five classes of fire. The type and amount of agent contained in the extinguisher and the extinguisher's design determine the amount of fire that can be extinguished for a particular class of fire. This information is conveyed by an alphanumeric classification system designed by Underwriters Laboratories Inc. (UL). NFPA 10 recommends that this rating information be displayed on the front faceplate of the extinguisher **(Figure 1.3, p. 11)**.

Table 1.1
Classification of Fire

Class Name	Letter Symbol	Image Symbol	Description
Class A or Ordinary Combustibles	**A** Ordinary Combustibles		Includes fuels such as wood, paper, plastic, rubber, and cloth
Class B or Flammable and Combustible Liquids and Gases	**B** Flammable Liquids		Includes all hydrocarbon and alcohol based liquids and gases that will support combustion.
Class C or Electrical	**C** Electrical Equipment		This includes all fires involving energized electrical equipment.
Class D or Combustible Metals	**D**		Examples of combustible metals are: magnesium, potassium, titanium, and zirconium.
Class K or Kitchen	**K**		Includes unsaturated cooking oils in well-insulated cooking appliances located in commercial kitchens.

Reproduced with permission from Wayne State University, Detroit, MI.

The rating system is based upon the extinguishment of test fires in accordance with UL 711, *Standard For Rating and Fire Testing Fire Extinguishers.* The ratings consist of both a numeral and a letter for extinguishers intended to combat Class A and Class B fires. Extinguishers classified for Class C fires receive only a letter rating because fires involving energized electrical equipment are fueled by materials that are typically Class A or Class B, or both, in composition. The primary concern in designating an extinguisher for Class C fires is to indicate to its user that it is safe against an electrical shock hazard. Likewise, Class D extinguishers receive only a letter designation. If an extinguisher is capable of extinguishing several classes of fire, the alphanumeric designation will denote this information by showing multiple number-letter ratings on the faceplate.

For example, on a 2A-10B:C rated fire extinguisher, the number 2 refers to the quantity of Class A material it is capable of extinguishing. The number 10 refers to the quantity of Class B material that it is capable of extinguishing. The numbers are determined after the rating tests are completed.

Figure 1.3 The faceplate of a fire extinguisher should clearly indicate the classes of fire the extinguisher is designed to handle.

There are several possible combinations of multipurpose extinguishers:

- A:B
- A:B:C
- A:C
- A: B: C: K
- B: C

NOTE: Class D ratings are never assigned to any type of multipurpose extinguisher.

> # WARNING!
> Never substitute an extinguishing agent inside an extinguisher for the agent indicated on the label of the extinguisher. For example, an extinguisher rated A: C is a water-mist variety, which uses only distilled water. The C rating is lost if ordinary tap water is substituted for distilled water and the extinguisher will be rendered ineffective.

General Rating Criteria

All extinguishers are tested in order to obtain a UL rating. In addition to the fire tests used to determine an extinguisher rating, an extinguisher must meet the following additional criteria:

- Volume of agent discharge
- Duration of agent discharge
- Range of discharge
- Hydrostatic testing of the agent vessel and discharge hose

Discharge Volume Capability

The following discharge capability characteristics are evaluated:

- Any portable fire extinguisher that uses dry chemical or dry powder as an extinguishing agent must be able to discharge 80 percent of its contents (**Figure 1.4**).
- Any portable fire extinguisher that uses agents other than dry chemical or dry powder (Halon, carbon dioxide, water, etc.) must be able to discharge 95 percent of its contents.

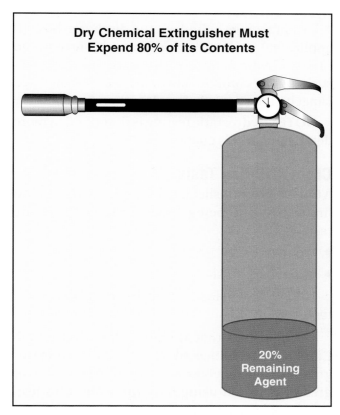

Dry Chemical Extinguisher Must Expend 80% of its Contents

20% Remaining Agent

Figure 1.4 To be effective, a dry chemical extinguisher must be able to expend 80 percent of its contents.

Discharge Duration

The following discharge duration criteria must be met:

- A portable extinguisher that uses water stored under pressure as an extinguishing agent must have a minimum 45- to 65-second discharge time. This time period will depend upon the desired rating.
- A minimum effective discharge time is required for any extinguisher for which a Class B rating is desired.

Discharge Range

The effective range of the extinguisher stream must meet the following requirements:

- Extinguishers that use water stored under pressure must have a minimum effective discharge of 30 feet (10 m) for a 40-second period.
- Dry chemical and dry powder extinguishers must have a minimum horizontal discharge range of 10 feet (3.1 m).

Hydrostatic Test

All agent storage cylinders, the discharge hose (if applicable), and the discharge nozzle are required to pass a hydrostatic test. This test consists of pressurizing the components to five times their rated capacity for a period of not less than 5 seconds.

NOTE: Hydrostatic testing is described in more detail later in this chapter.

Class A Rating Tests

A Class A extinguisher is subjected to two standard fire tests before being issued a rating. These fire tests are:

- Wood crib test
- Wood panel test

NOTE: Excelsior is used in the wood panel test.

Extinguishers that are rated Class 1-A through Class 6-A are subjected to both tests. An extinguisher that receives a rating of Class 10-A or greater is tested using only the wood crib test. Each type of test fire is unique with respect to the configuration and amount of Class A combustibles an extinguisher must extinguish before receiving its rating.

NOTE: Descriptions of extinguisher rating tests can be found in **Appendix B, Extinguisher Rating Tests**.

Class B Rating Tests

Like Class A extinguishers, extinguishers used for combating Class B fires are classified with a numerical designation. The number is an indication of the approximate area of a fire involving a 2-inch (50 mm) layer of flammable liquid that can be extinguished by a novice or inexperienced operator. The rating is based on the principle that an expert extinguisher operator such as a laboratory technician can extinguish two and one-half times more fire than a novice. For example, a novice extinguisher operator using a 60-B rated extinguisher can be expected to extinguish a flammable liquid fire involving a 60 square foot (5.6 m²) area. An expert using an extinguisher with the same rating, however, should be able to extinguish a fire involving an area of 150 square feet (14 m²). Ratings of 20-B or greater are considered for outdoor fires. **Table 1.2** shows the test criteria for various extinguisher ratings.

Class C Rating Tests

A Class C rating is not assigned a numerical designation because the rating signifies only that the extinguishing agent is electrically nonconductive (dielectric). The rating is provided in conjunction with a Class A or Class B rating (for example, 2-A, 10-B:C). No effort is made to indicate the extinguisher's capacity to extinguish a fire that includes energized electrical equipment. This is because fires involving energized electrical equipment are fueled by materials that are Class A or Class B in nature. For example, in a Class C fire, it may actually be the insulation material that is burning. The criteria for the rating of a Class C extinguisher are established by measuring the electrical conductivity of the extinguisher when it is discharged at an electrically energized target **(Figure 1.5, p. 14)**. It is equally important that the extinguishing agent be nonconductive in order to protect the operator.

Class D Rating Tests

Class D fire extinguishers are not given numerical ratings **(Figure 1.6, p. 14)**. Class D fire extinguishers are generally tested against fires involving specific combustible metals including magnesium, sodium, and potassium. Other tests, which are different in nature from the standard tests, may be

Rating ± Class	Minimum effective discharge time, seconds	Pan size (inside)		Metal thickness		Reinforcing angle size (approximate)		Commercial Grade heptane used, (approximate)[a]	
		mm²	(ft²)	mm	(in)	mm	(in)	Liters	(U.S. gallons)
Indoor Tests:									
1-B	8	0.23	2½	6	(¼)	38 by 38 by 4.8	(1½ by 1½ by 3⁄16)	11	(3-¼)
2-B	8	0.47	(5)	6	(¼)	38 by 38 by 4.8	(1½ by 1½ by 3⁄16)	23	(6-¼)
5-B	8	1.16	(12½)	6	(¼)	38 by 38 by 4.8	(1½ by 1½ by 3⁄16)	57	(15½)
10-B	8	2.32	(25)	6	(¼)	38 by 38 by 4.8	(1½ by 1½ by 3⁄16)	114	(31)
20-B	8	4.65	(50)	6	(¼)	38 by 38 by 4.8	(1 ½ by 1 ½ by 3⁄16)	227	(65)
Outdoor Tests:									
30-B	11	7.00	(75)	12	(½)	38 by 38 by 6	(1½ by 1½ by ¼)	350	(95)
40-B	13	9.30	(100)	12	(½)	38 by 38 by 6	(1½ by 1½ by ¼)	465	(125)
60-B	17	14.00	(150)	12	(½)	38 by 38 by 6	(1½ by 1½ by ¼)	700	(190)
80-B	20	18.60	(200)	12	(½)	38 by 38 by 6	(1½ by 1½ by ¼)	950	(250)
120-B	26	27.9	(300)	12	(½)	38 by 38 by 6	(1½ by 1½ by ¼)	1400	(375)
160-B	31	37.2	(400)	12	(½)	38 by 38 by 6	(1½ by 1½ by ¼)	1850	(500)
240-B	40	55.7	(600)	12	(½)	38 by 38 by 6	(1½ by 1½ by ¼)	2800	(750)
320-B	48	74.3	(800)	12	(½)	38 by 38 by 6	(1½ by 1½ by ¼)	3700	(1000)
480-B	63	112.0	(1200)	12	(½)	38 by 38 by 6	(1½ by 1½ by ¼)	5600	(1500)
640-B	75	149.0	(1600)	12	(½)	38 by 38 by 6	(1½ by 1½ by ¼)	7500	(2000)

[a] The amount of commercial grade heptane to be used in each test is to be determined by the actual depth as measured in the pan and not by the volume indicated.

Reprinted with permission from UL 711, *Rating and Fire Testing of Fire Extinguishers*, Seventh Edition, 2004. This is dated material. Copies of the current standard can be obtained from Underwriters Laboratories Inc., 333 Pfingsten Rd., Northbrook, IL 60062. Attn: Publications Stock.

UL shall not be responsible to anyone for the use of or reliance upon a UL Standard by anyone. UL shall not incur any obligation or liability for damages, including consequential damages, arising out of or in connection with the use, interpretation of, or reliance upon a UL Standard.

This information is reproduced/provided, with permission, from Underwriters Laboratories Inc. and is copyrighted by Underwriters.

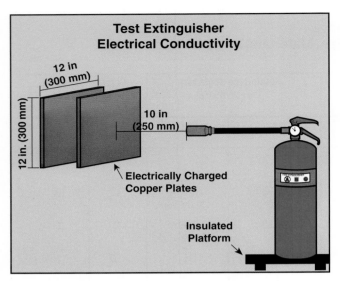

Test Extinguisher Electrical Conductivity

12 in (300 mm)

12 in. (300 mm)

10 in (250 mm)

Electrically Charged Copper Plates

Insulated Platform

Figure 1.5 A diagram of the test layout for Class C extinguishers.

Figure 1.6 Class D extinguishers do not receive a numerical rating.

required to cover situations or metals where the manufacturer's recommendations or the intended use of the extinguisher is indicated.

In addition to the fire tests, each extinguishing agent is evaluated with respect to the adverse effects that might occur in the course of discharge. These effects include agent toxicity, fumes developed, and products of combustion. The possibility of adverse reactions resulting from mistaken use of the agent on a combustible metal are also evaluated.

Depending on the intended use of the extinguisher, there are six basic fire tests that are performed. These include four tests involving magnesium and two tests involving sodium and potassium. Each test is intended to represent a typical fire situation in which a combustible metal might be involved.

Magnesium Fire Tests
There are four magnesium fire tests that are used to evaluate Class D portable fire extinguishers:

- Area fire test
- Pallet transfer fire test
- Premix fire test
- Casting fire test

Sodium and Potassium Fire Tests
UL Standard 711, *Rating and Testing of Fire Extinguishers,* specifies two standard procedures to test the ability of an agent to control fires involving sodium, potassium, and sodium-potassium alloys. These procedures are a spill fire test and a pan fire test. Both tests are conducted when the metals are in the liquid state.

Class K Rating Tests
Class K rated fire extinguishers are rated for their ability to fight fire in commercial cooking environments. The newest version of the UL Standard 300, *Standard for Fire Testing of Fire Extinguishing Systems for Protection of Restaurant Cooking Areas,* is more restrictive than previous tests in commercial cooking media. The fuel used must be new vegetable shortening or oil with an antifoaming agent and an auto-ignition temperature

of 685° F (362.7° C) or higher. Tests are performed on four types of cooking appliances: fryers, ranges, griddles, and woks.

Extinguishing Agents

Portable fire extinguishers use many different extinguishing agents. Each agent may be able to control one or more classes of fire, but one agent cannot extinguish all classes of fire. The following sections highlight the more common extinguishing agents.

Water

Water extinguishes primarily by cooling the burning fuel. Water is inexpensive and readily available, and water extinguishers – both plain and distilled – are relatively easy to maintain. Water does have some limitations as an extinguishing agent, however. Water is only effective on Class A fires; plain water, in itself, is ineffective on most Class B fires. Because water conducts electricity, it also should not be used on Class C fires. Water extinguishers are subject to freezing and therefore should be kept in an indoor area. Furthermore, the fire fighting capability of a portable water extinguisher is limited by the amount of water that the operator can easily carry. At nearly 42 pounds (19 kg), a 5-gallon (20 L) capacity water extinguisher is unwieldy and heavy to transport. For these reasons, a 5-gallon water extinguisher is the maximum size that can be considered to be portable **(Figure 1.7)**.

NOTE: Water extinguishers are also commonly available in a smaller 2½-gallon (10 L) size.

Relatively new to the realm of water extinguishing agents is the concept of using distilled water. The mineral content in plain water enhances its ability to conduct electricity, but distilled water has had most of its mineral content removed. The absence of mineral content, plus the use of an atomizing applicator, makes distilled water useful as an extinguishing agent on electrical fires. The atomizing applicator discharges the agent in a very fine mist, which also contributes to its nonconductive characteristic. Underwriters Laboratories Inc. has rated the 2½-gallon (10 L) distilled water fire extinguisher at 2 A:C.

Figure 1.7 Water-filled extinguishers are heavy. A 5-gallon (20 L) size is the largest size that is considered to be portable. *Courtesy of Tyco Safety Products.*

Antifreeze

Antifreeze agents are used in portable fire extinguishers designed and tested for operation in temperatures as low as -40°F (-40°C). Antifreeze agents are intended for use on Class A fires in locations where low temperatures make other water-type extinguishers unusable. The fire extinguishing rating tests are developed at normal ambient temperatures (approximately 70° F [21° C]) and effectiveness may be reduced at lower temperatures. The extinguishing agents are alkali-metal salt solutions or calcium chloride. Care must be taken to follow charging and maintenance directions on the extinguisher nameplate or recharging package.

Alkaline Mixtures

Alkaline mixtures have been found to be effective against fires originating in cooking-related media. These mixtures consist of potassium acetate, potassium citrate, potassium carbonate, or combinations thereof, in water. Alkaline mixtures are particularly useful for attacking a Class K fire because of their ability to generate soapy foam. The generation of this soapy foam is called *saponification* and occurs as a result of reactions with fats. Saponification generates steam (a cooling effect), smothers the fire, and prevents reignition. A:B:C or multipurpose dry chemical extinguishing agents do not produce the saponification effect because the base chemical – ammonium phosphate – is acidic in nature. Recent tests have shown that fire losses are lower, extinguishment is faster and safer, and cleanup is minimal with an alkaline-type extinguishing agent. In addition to a Class K rating, most of the extinguishers of this type are given a Class A, Class B, and Class C rating.

Carbon Dioxide

Carbon dioxide (CO_2) is a colorless, noncombustible gas that is heavier than air. It extinguishes fire primarily through a smothering action by establishing a gaseous blanket between the fuel and the surrounding air. Carbon dioxide is suitable for Class B and Class C fires (**Figure 1.8**). It has very limited value on deep-seated Class A fires, which can rekindle after the carbon dioxide dissipates into the atmosphere. Because of carbon dioxide's gaseous nature, it is difficult to project it very far from the discharge horn of the extinguisher.

The carbon dioxide is stored in the extinguisher in a liquid state at a pressure of about 840 psig (5 880 kPa). Storing the carbon dioxide as a liquid allows more weight to be stored in a given volume. When discharged from the extinguisher, the carbon dioxide has a white cloudy appearance. This is due to the small dry ice crystals that may be carried along in the gas stream when it is discharged. Carbon dioxide extinguishers characteristically discharge with a loud noise that may startle an untrained operator.

> ### WARNING!
> Carbon dioxide is an asphyxiant. Be very cautious about using it in a small or confined space.

Aqueous Film Forming Foam (AFFF)

Aqueous film forming foam (AFFF) produces both air foam and a floating film on the surface of a liquid fuel (**Figure 1.9**). AFFF is suitable for both Class A and Class B fires. Most commonly, the AFFF concentrate is premixed with water in the extinguisher and discharged through a special nozzle. Because the foam agent is mixed with water, AFFF is effec-

Figure 1.8 Because carbon dioxide extinguishers work by smothering the fire, they must be used with caution in small or enclosed spaces.

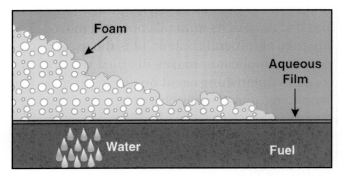

Figure 1.9 AFFF creates a foam that floats on the surface of the fuel.

tive on Class A fires by cooling and penetrating the fuel. This agent is very effective on flammable liquid fires because of the double effect of a foam blanket and a surface film to exclude air from the fuel. The AFFF/water mix has all the same inherent limitations discussed earlier for plain water (**Figure 1.10**).

///////////////////////////////////

CAUTION!

Foam extinguishers are only useful on relatively small flammable liquid fires. Foam does not provide much radiant heat protection, so the extinguisher operator is at serious risk of injury if the fire is too big to be handled by the extinguisher. A dry chemical extinguisher would be a much better choice in that situation.

Another type of AFFF extinguisher currently available contains plain water in the vessel and a special nozzle with a solid form of AFFF concen-

trate. As the water flows through the nozzle, the concentrate is dissolved in the water to produce a finished AFFF solution.

Film Forming Fluoroprotein (FFFP)

Film forming fluoroprotein (FFFP) is a foaming agent very similar to that of AFFF. This foam is usually diluted in water to a 3 or 6 percent solution and is effective on Class A and Class B fires (**Figure 1.11**). A typical FFFP portable fire extinguisher is found in a 2½-gallon size and rated 3A:20B. Expect to find these extinguishers in areas where gasohol and water-soluble flammable liquids are in use. For

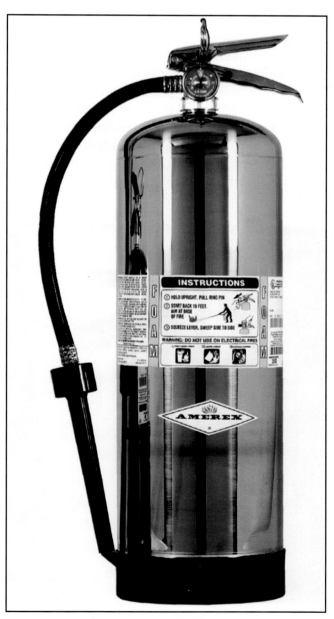

Figure 1.11 An FFFP extinguisher. *Courtesy of Amerex.*

Figure 1.10 A wheeled AFFF extinguisher. *Courtesy of Tyco Safety Products.*

more detailed information on the FFFP agent and other foams, see the IFSTA **Principles of Foam Fire Fighting** and **Fire Inspection and Code Enforcement** manuals.

NOTE: The way in which extinguishers are rated is covered later in this chapter.

Halons

Halons and the halogenated extinguishing agents contain atoms from one of the halogen series of chemical elements: fluorine, chlorine, bromine, and iodine. The halogenated agents are principally effective on Class B and Class C fires **(Figure 1.12)**. Halons were originally developed and used because they were considered a "clean agent," namely one that leaves no residue. Because Halon has been proven to be harmful to humans and to the earth's ozone layer, however, international restrictions have been placed on its production. Although the Montreal Protocol of 1987 provided for a phase-out of halons and forbade the manufacture of new Halon agents after January 1, 1994, limited production continues because there are some exceptions to the phase-out plan.

Figure 1.12 Several types of Halon extinguishers are still in use.

Use of Halons

Locations where Halon agent use is deemed to be "essential" may be granted an exemption from the phase-out, and Halon systems installed prior to the Montreal Protocol may remain in use until such time as they are discharged on a fire or the gas "leaks off." The criteria for this exemption are as follows:

- Halon agent use is necessary for human health and safety or critical for the functioning of society.
- There are no technically or economically feasible alternatives.
- All feasible actions must be taken to minimize emissions from use.
- The supply of Halon agents from existing banks or recycled stocks is not sufficient to accommodate the need.

In portable fire extinguishers, two Halons are still in use:

- Halon 1211 (bromochlorodifluoromethane — CF_2BrCl)
- Halon 1301 (bromotrifluoromethane — CF_3Br)

Halon 1211 is the one most commonly found in portable fire extinguishers. Halon 1301 is used in some portable fire extinguishers, but it is more commonly found in fixed system applications.

CAUTION!

Avoid prolonged exposure to the Halons. These substances break down under fire conditions and liberate toxic substances such as chlorine, bromine, hydrogen chloride, and hydrogen bromide. These are irritating gases and their presence is readily apparent to the extinguisher user.

Halon Replacement Agents

There has been considerable research and development on new clean agents that extinguish fires in the same manner as Halon agents, but that cause no significant damage to the atmosphere. NFPA 2001, *Standard on Clean Agent Fire Extinguishing*

Systems (2004) defines a clean agent as an "electrically nonconducting, volatile, or gaseous fire extinguishant that does not leave a residue upon evaporation."

Several categories of clean agents are now commercially available: halocarbon clean agents and inert gas clean agents. Halocarbon clean agents include hydrochlorofluorocarbon (HCFC) and hydrofluorocarbon (HFC). Common Halon replacement agents include Halotron®, FM-200, and Inergen.

NOTE: State and local codes may vary with respect to conversion to Halon replacement agents, which cannot usually be put directly into an existing Halon extinguishing system or portable extinguisher without certain precautions. For more information, see NFPA 2001.

Halotron®

Halotron® is a "clean agent" hydrochlorofluorocarbon which, when discharged, is a rapidly evaporating liquid. Halotron® leaves no residue and meets Environmental Protection Agency (EPA) minimum standards for discharge into the atmosphere. The agent does not conduct electricity back to the operator, making it suitable for Class C fires. Halotron® has a limited Class A rating for extinguishers over 9 pounds (4 kg). A 28-pound (13 kg) Halotron® extinguisher is given a UL rating of 2A:10BC. It may be found in telecommunication facilities, clean rooms, computer rooms, and even vehicles **(Figure 1.13)**.

FM-200

FM-200 is a hydrofluorocarbon that is considered to be an acceptable alternative to Halon 1301 because it leaves no residue and is not harmful to humans or the environment. This agent does require significantly more agent to achieve extinguishment than Halon 1309 did.

Inergen

Inergen is a blend of three naturally occurring gases: nitrogen, argon, and carbon dioxide. It is stored in cylinders near the facility under protection. Inergen is environmentally safe and does not contain a chemical composition like many other proposed Halon alternatives.

Figure 1.13 Halotron® is one of the "clean agents" used to replace Halon extinguishers. *Courtesy of Ted Boothroyd.*

NOTE: A list of additional clean agents can be found in **Table 1.3, p. 20.** Consult NFPA 2001, *Clean Agent Fire Extinguishing Systems*, 2004 Ed. for more information.

Dry Chemical Agents

Several dry chemicals have proven useful as extinguishing agents in portable fire extinguishers. These agents, called "ordinary" dry chemical agents, include sodium bicarbonate, potassium bicarbonate, urea bicarbonate, and potassium chloride. In addition, there is monoammonium phosphate, a "multipurpose" dry chemical.

In physical form, the dry chemicals are very small solid powdery particles. Because these particles are solid, they can be projected more effectively from the extinguisher nozzle than gaseous agents **(Figure 1.14, p. 21)**. Therefore, dry chemicals do not dissipate into the atmosphere as readily as gases and are especially suitable for controlling

Table 1.3
Clean Agents

FC-3-1-10	Perfluorobutane	C_4F_{10}
FK-5-1-12	Dodecaflouro-2	$CF_2CF_2C(O)CF(CF_3)_2$
	Methylpentan-3-one	
HCFC Blend	Dichlorotrifluoroethane	$CHCl_2CF_3$
A	HCFC-123 (4.75%)	
	Chlorodifluoromethane	$CHClF_2$
	HCFC-22 (82%)	
	Chlorotetrafluoroethane	$CHClFCF_3$
	HCFC-124 (9.5%)	
	Isopropenyl-1-	
	Methylcyclohexene (3.75%)	
HCFC-124	Chlorotetrafluorethane	$CHClFCF_3$
HFC-125	Pentafluoroethane	CHF_2CF_3
HFC-227ea	Heptafluoropropane	CF_3CHFCF_3
HFC-23	Trifluoromethane	CHF_3
HFC-236fa	Hexafluoropropane	$CF_3CH_2 CF_3$
FIC-13I1	Trifluoroiodide	CF_3I
IG-01	Argon	Ar
IG-100	Nitrogen	N_2
IG-541	Nitrogen (40%)	N_2
	Argon (40%)	Ar
	Carbon Dioxide (8%)	CO_2
IG-55	Nitrogen (50%)	N_2
	Argon (50%)	Ar

Notes:

1. Other agents could become available at later dates. They could be added via the NFPA process in future additions or amendments of the standard.

2. Compositton of inert gas agents are given in percent by volume. Composition of HCFC Blend A is given in percent by weight.

3. The full analogous ASHRAE nomenclature for FK-5-1-12 is FK-5-1-12mmy2.

Figure 1.14 Dry chemical agent can be projected farther than gaseous agents. *Courtesy of Tyco Safety Products.*

toxic effects but may be slightly irritating. Sodium bicarbonate is color-coded either blue or white to distinguish it from other dry chemical agents.

Sodium bicarbonate is effective on Class B and Class C fires. It was formerly used widely for protection of commercial food preparation equipment such as fryers and range hoods. Although still effective for use in residential kitchens, sodium bicarbonate is no longer recognized by UL300 to be used as an extinguishing agent for commercial kitchen fires, now known as Class K fires. Manufacturers of hood extinguishing systems are now using the potassium-based liquid agents for installation in commercial kitchens.

Although sodium bicarbonate is no longer used for commercial kitchen fires, it is effective against other Class B fires. In fact, when evaluated against an equal weight of carbon dioxide, sodium bicarbonate is twice as effective against Class B fires. Sodium bicarbonate has a very rapid knockdown capability against flaming combustion and also has some effect on surface fires in Class A materials. In this connection, it has been used successfully on textile machinery where the fine textile fibers can produce a surface fire.

Potassium Bicarbonate

Potassium bicarbonate, also known as Purple K, has properties and applications similar to sodium bicarbonate. However, on a pound-for-pound basis, it is about twice as effective as sodium bicarbonate. That is, given a specific amount of potassium bicarbonate, an operator can extinguish a fire twice as large as one that can be extinguished with the same amount of sodium bicarbonate. Like sodium bicarbonate, potassium bicarbonate is most effective on Class B and Class C fires and is also treated to be water repellant and free flowing. Potassium bicarbonate is color-coded violet to differentiate it from other dry chemicals.

Monoammonium Phosphate

Monoammonium phosphate is effective on Class A, Class B, and Class C fires. It has an action similar to other dry chemicals on flammable liquid fires (**Figure 1.15, p. 22**). Using a combination of extinguishing methods, it quickly knocks down flaming combustion. On Class A materials, the monoam-

outdoor fires. The primary disadvantages of using dry chemical fire extinguishers are difficulties in cleanup. Airborne dry powder travels much farther than the immediate fire area, so cleanup can be very time-consuming.

As with all fire extinguishers, a dry chemical type should never be refilled with an agent other than the specific agent for which the extinguisher was designed. In addition, dry chemical extinguishing agents should never be mixed. This is especially true of "ordinary" dry chemicals and "multipurpose" dry chemicals. The mixture of monoammonium phosphate and other dry chemicals can result in a dangerous chemical reaction.

Sodium Bicarbonate

Sodium bicarbonate, a form of baking soda, was the first commercially produced dry chemical agent and is still widely used. The agent is treated to be water repellant and free flowing. When used as an extinguishing agent, sodium bicarbonate has no

monium phosphate forms a solid coating and extinguishes the fire by a smothering action. Once known as the ideal extinguishing agent, monoammonium phosphate is soon to lose its "ideal" status in favor of the potassium-based Class A-, B-, C-, and K-rated fire extinguishing agents.

Class D Dry Powders

Dry powder extinguishing agents are designed to extinguish Class D fires in combustible metals such as aluminum, magnesium, sodium, and potassium **(Figure 1.16)**. Ordinary extinguishing agents are not capable of controlling fires in combustible metals. Burning magnesium can break down water and water-based extinguishing agents. Violent reactions may occur, and large quantities of hydrogen gas are released when water contacts the burning metal. Some of the combustible metals can continue to burn in the inert gas agents. The

Halons can react violently with combustible metals and may break down into highly toxic gases.

A number of extinguishing agents have been developed for combustible metals, including somewhat exotic and specialized materials such as foundry flux, trimethoxyboroxine (in a liquid form), ternary eutetic chloride, and boron trifluoride (a gas). **There is no single agent that is effective on all combustible metals**. In a given situation, the extinguishing agent must be carefully chosen for the hazard being protected. The following sections address just three of the more commonly encountered Class D agents.

NA-X®. NA-X® is a Class D extinguishing agent designed specifically for use on sodium, potassium, and sodium-potassium alloy fires. NA-X® is not suitable for use on magnesium fires. Chemically, NA-X® has a sodium carbonate base combined with additives to enhance flow. The extinguishing action forms a crusting or caking on the burning

Figure 1.15 Monoammonium phosphate extinguishers are effective on many types of Class B fires.

Figure 1.16 Because ordinary extinguishing agents are not effective on combustible metals, dry powders have been developed for these hazards.

material, causing an oxygen deficiency and thereby extinguishing the fire. Underwriters Laboratories Inc. lists NA-X® for use on the burning materials at fuel temperatures up to 1,400° F (760° C). Application can be from portable extinguishers or by scoops from pails.

Met-L-X®. Met-L-X® is a sodium chloride (salt) based extinguishing agent intended for use on magnesium, sodium, and potassium fires. Like other dry powders, it contains additives to enhance flowing and prevent caking in the extinguisher. It also extinguishes metal fires by forming a crust on the burning metal to exclude oxygen. The agent is applied from the extinguisher to first control the fire, and then the agent is applied more slowly to bury the fuel in a layer of powder. The agent is stable when stored in sealed containers; it is nonabrasive and has no known toxic effects.

Lith-X®. Lith-X® is an agent that can be used on several combustible metals. It was developed to control fires involving lithium but can also be used to extinguish magnesium, zirconium, and sodium fires. Lith-X® consists of a graphite base that extinguishes fires by conducting heat away from the fuel after a layer of the powder has been applied to the fuel. Unlike other dry powders, it does not form a crust on the burning material.

Types of Fire Extinguishers

Portable fire extinguishers use different methods to expel the extinguishing agent and can be broadly classified according to the method used. These include the following:

• Stored-pressure
• Cartridge-operated
• Pump-operated

Stored-Pressure Extinguishers

A stored-pressure fire extinguisher contains an expellant gas and an extinguishing agent in a single chamber (**Figure 1.17**). The pressure of the gas forces the agent out through a siphon tube, valve, and nozzle assembly. The pressurizing gas can be a different gas than the extinguishing agent. For example, dry-chemical extinguishers typically use nitrogen as an expellant gas. In other cases, the expellant gas can be the vapor phase of the agent itself, such as that in carbon dioxide extinguishers.

Figure 1.17 A typical stored-pressure extinguisher. *Courtesy of Tyco Safety Products.*

As a highly compressed gas, carbon dioxide forms its own expellant. Units that use a separate expelling gas have a pressure gauge that permits visual determination as to whether the extinguisher is ready for use. Air pressurizing water (APW) extinguishers are among the most common type of stored-pressure extinguishers.

The stored-pressure extinguisher is simple to use. It usually requires only that the operator remove a safety pin and squeeze the valve handle. However, refilling the unit requires special charging equipment for pressurization, so service is usually performed by licensed distributors. This type of extinguisher may be found in such areas as office buildings, department stores, or even private residences where a high-use factor is not involved.

Cartridge-Operated Extinguishers

The cartridge-operated extinguisher has the expellant gas stored in a separate cartridge, while the extinguishing agent is contained in an adjacent

Cartridge

Figure 1.18 Note the separate cartridge used to store the extinguishing agent in a cartridge-operated extinguisher. *Courtesy of Tyco Safety Products.*

Figure 1.19 The advantage of a pump-operated extinguisher is that it can be refilled from any available water source. *Courtesy of Ted Boothroyd.*

cylinder, called an agent cylinder or tank (**Figure 1.18**). To actuate the extinguisher, the expellant gas (carbon dioxide or nitrogen) is released into the agent cylinder. The pressure of the gas forces the agent into the application hose. Discharge is controlled by a hand-held nozzle/lever. No pressure gauge is provided. During inspection, the expellant gas cartridge is weighed to ensure that it has adequate gas. Replacing the gas cartridge and filling the agent cylinder recharges this type of extinguisher. This procedure may be performed in-house and does not require special equipment. These extinguishers are found in industrial operations such as paint spraying or solvent manufacturing facilities, where they may be used frequently.

Pump-Operated Extinguishers

A pump-operated extinguisher discharges its agent by the manual operation of a pump (**Figure 1.19**). This type of extinguisher is limited to the use of water as the extinguishing agent. Its primary advantage is that it can be refilled from any avail-

able water source in the course of extinguishing a fire. Maintenance is extremely simple, consisting mainly of ensuring that the extinguisher is full and has not suffered any mechanical damage.

Selection and Distribution of Extinguishers

Extinguishers must be properly distributed throughout an occupancy to ensure that they are readily available during an emergency. To be effective, extinguishers cannot be located at great travel distances from where they may be needed. In the same manner, they cannot be effective if there are not enough extinguishers provided for the hazard involved. The proper selection and distribution of extinguishers is determined by several factors, including the size of the extinguisher and the hazard protected. Requirements for extinguisher distribution are contained in NFPA 10; these requirements are separated into specifics for Class A, Class B, Class C, Class D, and Class K hazards. Because local

Obsolete Extinguishers

Fire protection personnel must be alert for fire extinguishers that are out of production and no longer suitable for use. Operating these extinguishers, even when used as directed, could result in injury or death to the user. Despite the fact that such extinguishers are no longer approved for use, they are still encountered. The more common types of obsolete fire extinguishers that are encountered are the inverting-type fire extinguishers and soldered or riveted shell soda-acid, chemical foam, cartridge-operated water extinguisher, and loaded stream extinguishers **(Figure 1.20)**. In 1982, OSHA decreed that all of these be removed from service for the following reasons:

• The extinguishing agents conduct electricity.

• The extinguisher cannot be turned off once it is activated.

• The extinguishing agent is more corrosive than water.

• The extinguisher is potentially dangerous to the operator during use.

If the discharge hose becomes blocked, it can build up pressures in excess of 300 psi (2 100 kPa), resulting in injury or death.

• The agent tank may have corroded over the years. This weakness may result in a violent failure when the tank becomes pressurized, also resulting in injury or death. **All obsolete extinguishers should be removed from service immediately and replaced with extinguishers that meet the requirements specified in NFPA 10.**

Figure 1.20 All soda-acid and inverting-type extinguishers have long been obsolete and are considered hazardous.

codes and ordinances can be more restrictive, they should be reviewed along with the requirements contained in NFPA 10. The following elements are important in the selection and distribution of fire extinguishers:

• Chemical and physical characteristics of the combustibles that might be ignited

• Potential severity (size, intensity, and rate of advancement) of any resulting fire

• Location of the extinguisher

• Effectiveness of the extinguisher for the hazard in question

• Personnel available to operate the extinguisher, including their physical abilities, emotional

characteristics, and any training they may have in the use of extinguishers

• Environmental conditions that may affect the use of the extinguisher (temperature, winds, presence of toxic gases or fumes)

• Any anticipated adverse chemical reactions between the extinguishing agent and the burning material

• Any health and occupational safety concerns such as exposure of the extinguisher operator to heat and products of combustion during fire fighting efforts

• Inspection and service required to maintain the extinguishers

In addition, the type, size, and number of extinguishers needed vary according to whether the occupancy is classified as light hazard, ordinary hazard, or extra hazard. Also considered are "Nature-of-the-Hazard" and "Size-of-the-Extinguisher" factors. NFPA 10 describes the need for extinguisher selection and distribution as follows:

Nature-of-the-Hazard Factor

In a given situation, the specific nature of the hazard dictates the type, size, number and distribution of extinguishers. However, there is a general method that is used to determine a satisfactory distribution of extinguishers in the vast majority of situations. This method, which is contained in NFPA 10 and is used nationally, consists of classifying an occupancy as light (low) hazard, ordinary (moderate) hazard, or extra (high) hazard. Fire extinguisher distribution is specified on the basis of that classification.

Size-of-the-Extinguisher Factor

For each occupancy classification, NFPA 10 recommends the minimum size extinguisher needed and the maximum area to be protected by an extinguisher. **Table 1.4** defines the maximum areas that can be adequately protected by Class A extinguishers of a given rating. In all occupancies, the maximum travel distance to an extinguisher for Class A hazards is 75 feet (25 m).

In applying the figures listed in **Table 1.4,** the floor area of a given occupancy is divided by the figure listed for maximum floor area per unit of "A" for a given size of extinguisher. This calculation yields the minimum number of extinguishers that must be provided. For example, take an ordinary hazard occupancy that contains 9,000 square feet (836 m^2) and 2-A rated extinguishers (a very common size) are used. Each 2-A extinguisher can protect 3,000 square feet (279 m^2) of ordinary hazards (2-A × 1,500 ft^2/A). Then, 9,000 square feet (836 m^2) divided by 3,000 square feet (279 m^2) yields a requirement of three extinguishers. These extinguishers are then distributed so that the travel distance to an extinguisher does not exceed 75 feet (25 m).

If a fire extinguisher with a rating of 6-A were to be used in the above example, only one extinguisher would be needed for 9,000 square feet

Table 1.4
Fire Extinguisher Size and Placement
for Class A Hazards

Criteria	Light (Low) Hazard Occupancy	Ordinary (Moderate) Hazard Occupancy	Extra (High) Hazard Occupancy
Minimum rated single extinguisher	2-A*	2-A*	4-A+
Maximum floor area per unit of A	3,000 ft^2	1,500 ft^2	1,000 ft^2
Maximum floor area for extinguisher	11,250 ft	11,250 ft	11,250 ft
Maximum travel distance to extinguisher	75 ft	75 ft	75 ft

For SI units: 1 ft = 0.305 m; 1 ft^2 = 0.0929 m^2
* Up to two water-type extinguishers, each with 1-A rating, can be used to fulfill the requirements of one 2-A rated extinguisher.
+ Two 2 ½-gal (9.46 L) water-type extinguishers can be used to fulfill the requirements of one 4-A rated extinguisher.

(836 m²). It is important to remember, however, that the travel distance still could not exceed 75 feet (25 m).

If the area to be protected is less than 3,000 square feet (279 m²) (6,000 square feet [558 m²] for light hazard), at least one extinguisher of a 2-A rating must be provided. Of course, more extinguishers than the minimum specified by the table can be provided.

Light-Hazard Occupancy

A light-hazard occupancy is one in which the amount of ordinary combustible material or flammable liquids present is such that a fire of small size may be expected. Such occupancies include classrooms (but not necessarily all parts of a school), churches, and assembly halls (**Figure 1.21**).

Ordinary-Hazard Occupancy

An ordinary-hazard occupancy is one in which the amount of ordinary combustibles and flammable liquids present would result in an incipient fire of moderate size. In such occupancies, fire growth would not be so rapid as to be beyond control by extinguishers if the fire were discovered quickly. Occupancies considered ordinary hazard include mercantile storage and display, light manufacturing facilities, and parking garages and warehouses not classified as extra hazard.

NOTE: Some parts of an occupancy that are classified as light hazard may actually be ordinary hazard. Examples of this situation are the shop or storage areas in a school.

Extra-Hazard Occupancy

Extra-hazard occupancies are those where the amounts of ordinary combustible materials and flammable liquids present are high and a rapidly spreading fire may develop. Examples of extra-hazard occupancies include automotive repair shops, painting facilities, manufacturing operations that use flammable liquids, restaurants with deep fat fryers, and locations with high-piled storage of combustibles (**Figures 1.22 a and b**).

Class A Extinguisher Distribution Factors

In ordinary- or low-hazard occupancies, the authority having jurisdiction may approve the use of several lower-rated extinguishers in place of

Figures 1.22 a and b Restaurant kitchens and paint storage areas are typical examples of extra-hazard occupancies.

Figure 1.21 Churches are typically light-hazard occupancies.

one higher-rated extinguisher. For example, two or more extinguishers may be used to fulfill a 6-A rating if there are enough individuals trained to use the extinguishers. When the weight of the extinguisher causes problems for those who will be operating it, two extinguishers of lesser weight may be used to replace the heavier extinguisher.

Class B Extinguisher Distribution Factors

Determining the distribution of Class B extinguishers depends upon the travel distance to the hazard. Flammable liquid fires develop very rapidly and occur in a variety of situations that are fundamentally different from a fire control standpoint. In providing extinguishers for Class B hazards, two situations may be encountered. One is a spill fire where the flammable liquid does not have depth, and the other involves flammable liquids with depth, such as dip tanks. NFPA 10 establishes ¼ inch (6 mm) deep as the criterion for a flammable liquid fire to be considered to be "with depth." Anything less than this is considered to be "without depth."

Flammable Liquid Fires Without Depth

In determining the proper distribution of extinguishers for flammable liquid fires without depth, the travel distances shown in **Table 1.5** are used. Flammable liquids without depth refers to those situations where the liquids are used in a process but are not found in large, deep, open containers. Notice that the table specifies a travel distance only; no area is specified. Due to the rapidly developing nature of a flammable liquid fire, the speed with which the operator can begin to use an extinguisher is extremely important. For this reason, the travel distance to a flammable liquid hazard is less than that permitted for Class A hazards.

In the case of flammable liquid fires, multiple extinguishers with lower ratings generally cannot be used to satisfy a requirement for a larger unit. The larger unit is necessary because of the possibility of a flashback over the surface of the liquid if the fire is not completely extinguished with a small extinguisher. An exception to this rule is made for AFFF extinguishers where up to three extinguishers can be used to satisfy requirements in low- and extra-hazard occupancies. A foam extinguisher can establish a blanket of agent on the surface of the liquid to eliminate a flashback.

Table 1.5 Fire Extinguisher Size and Placement for Class B Hazards

Type of Hazard	Basic Minimum Extinguisher Rating	Maximum Travel Distance to Extinguishers
Light (low)	5-B	30 feet (9.15 m)
	10-B	50 feet (15.25 m)
Ordinary (moderate)	10-B	30 feet (9.15 m)
	20-B	50 feet (15.25 m)
Extra (high)	40-B	30 feet (9.15 m)
	80-B	50 feet (15.25 m)

Notes:
(1) The specified ratings to not imply that fires of the magnitudes indicated by these ratings will occur, but rather they are provided to give the operators more time and agent to handle difficult spill fires that could occur.
(2) For fires involving water-soluble flammable liquids, see 4.3.4.
(3) For specific hazard applications, see Section 4.3.

Reprinted with permission from NFPA 10, *Standard for Portable Fire Extinguishers,* Copyright© 2002 National Fire Protection Association, Quincy, MA 02269. This reprinted material is not the complete and official position of the National Fire Protection Association on the referenced subject, which is represented only by the standard in its entirety.

CAUTION!

In general, multiple extinguishers with lower ratings cannot be use to satisfy a requirement for a larger unit. Flammable liquids are capable of flashing back and injuring the extinguisher operator if the fire is not extinguished quickly.

Flammable Liquid Fires with Depth

Individual hazards involving flammable liquids with depth are often protected by fixed extinguishing systems. These systems lessen the requirements for portable fire extinguishers in the area but do not eliminate the need for them. A spill fire could occur beyond the effective reach of a fixed system and portable extinguishers would be needed.

If there is no fixed system, an extinguisher must be provided that has a numerical classification equal to twice the surface area of the largest hazard in the occupancy. For example, an extinguisher with a 10-B rating would be needed to protect 5 square feet (.5 m²). It is possible to specify extinguishers in this manner because the surface area is usually known when flammable liquids are present in depth. An example would be an industrial occupancy with dip tanks or retention dikes. As with surface fires, smaller extinguishers cannot be used instead of a required larger extinguisher, although up to three AFFF extinguishers may be used to satisfy requirements just as with surface fires.

Extinguishers for protection of flammable liquid hazards must be placed so that an operator is not endangered while attempting to reach an extinguisher. Extinguishers should not be placed over or behind a hazard.

Class C and Class D Extinguisher Distribution Factors

There are no special spacing rules for Class C hazards because fires involving energized electrical equipment usually involve Class A or Class B fuels. Furthermore, the placement and distribution of fire extinguishers for Class D combustible metals cannot be generalized. Determining extinguisher placement involves making an analysis of the specific metal, the amount of metal present, the configuration of the metal (solid or particulate), and the characteristics of the extinguishing agent. NFPA 10 recommends only that the travel distance for Class D extinguishers not exceed 75 feet (25 m).

Class K Extinguisher Distribution Factors

In the working environment of commercial cooking occupancies, fire is always present. Employees in such areas are charged with the responsibility to maintain appropriate cooking temperatures to ensure safety. Because employees are in various levels of training for their jobs and because there is the potential of fire hazards occurring in an assembly area (dining room), NFPA 10 has assigned a more restrictive distance requirement. In areas where Class K fires are likely, the maximum travel distance from the hazard to the extinguisher is reduced to 30 feet (10 m).

Installation and Placement of Extinguishers

In addition to proper selection and distribution, effective use of fire extinguishers requires that they be readily visible and accessible. Proper extinguisher placement is an essential but often overlooked aspect of fire protection. Extinguishers should be mounted properly to avoid injury to building occupants and to avoid damage to the extinguisher. Some examples of improper mounting would be an extinguisher mounted where it protrudes into a path of travel or one that is sitting on top of a workbench with no mount at all. To minimize these problems, extinguishers are frequently placed in cabinets or wall recesses for protection of both the extinguisher and people who might walk into them **(Figure 1.23)**. If an extinguisher cabinet is placed in a rated wall, then the cabinet must have the same fire rating as the wall assembly. Proper placement of extinguishers should provide for the following:

Figure 1.23 Placing an extinguisher in a cabinet helps protect both the extinguisher and passersby who might walk into it.

- Extinguishers should be visible and well signed.

- Extinguishers should not be blocked by storage or equipment.

- Extinguishers should be near points of egress or ingress.

- Extinguishers should be near normal paths of travel.

Although an extinguisher must be properly mounted, it must be placed so that all personnel can access it. The extinguisher should not be placed too high above the floor for safe lifting. The standard mounting heights specified for extinguishers are as follows:

- Extinguishers with a gross weight not exceeding 40 pounds (18 kg) should be installed so that the top of the extinguisher is not more than 5 feet (1.5 m) above the floor **(Figure 1.24)**.

- Extinguishers with a gross weight greater than 40 pounds (18 kg), except wheeled types, should be installed so that the top of the extinguisher is not more than 3½ feet (1 m) above the floor.

- The clearance between the bottom of the extinguisher and the floor should never be less than 4 inches (100 mm).

Physical environment is very important to extinguisher reliability. The greatest concern is the temperature of the environment. Because testing laboratories evaluate water-based extinguishers at temperatures between 40° F and 120° F (4° C and 49° C), these extinguishers must be located where freezing is not possible. Other types of extinguishers can be installed where the temperature is as low as –40° F (–40° C). Specialized extinguishers are available for temperatures as low as –65° F (–54° C). Extinguishers using plain water can be provided with antifreeze recommended by the manufacturer. Care must be exercised in the use of antifreeze, however. Ethylene glycol cannot be used, and calcium chloride cannot be used in stainless steel units. Antifreeze cannot be added to AFFF extinguishers.

Other environmental factors that may adversely affect an extinguisher's effectiveness are snow, rain, and corrosive fumes. A corrosive atmosphere can be encountered not only in an industrial environment but also in marine applications where

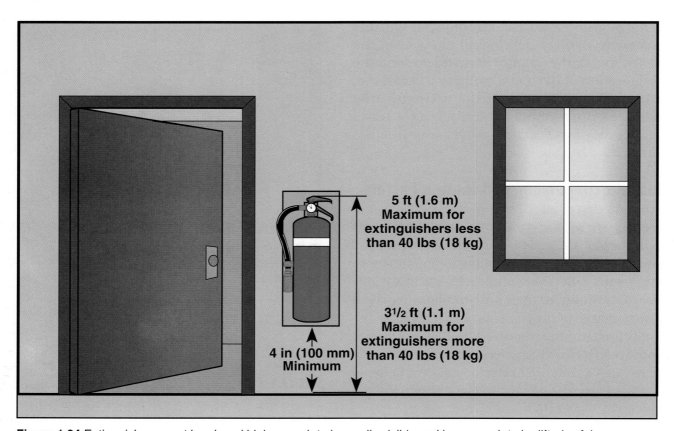

5 ft (1.6 m) Maximum for extinguishers less than 40 lbs (18 kg)

3½ ft (1.1 m) Maximum for extinguishers more than 40 lbs (18 kg)

4 in (100 mm) Minimum

Figure 1.24 Extinguishers must be placed high enough to be easily visible and low enough to be lifted safely.

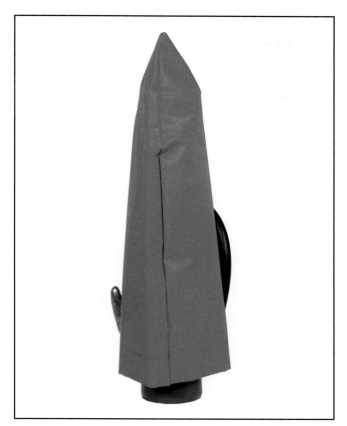

Figure 1.25 An extinguisher may need a protective covering to avoid damage and corrosion. *Courtesy of Tyco Safety Products.*

Figure 1.26 Extinguishers carried on fire apparatus receive a lot of wear and tear. *Courtesy of Tyco Safety Products.*

extinguishers are exposed to saltwater spray. In the case of outdoor installations, the extinguisher can be protected with a plastic bag or placed in a cabinet **(Figure 1.25)**. For marine applications, extinguishers are available that have been listed for use in a saltwater environment.

Portable Fire Extinguishers on Fire Apparatus

Portable extinguisher use is not limited to buildings or other structures. In fact, in the hands of a trained firefighter, the extinguisher probably attains its maximum effectiveness. Most of the same considerations that apply to the use of extinguishers by the general public also apply to their use by fire department personnel. Extinguishers used by firefighters must be the proper type, readily accessible, and properly maintained.

Extinguishers carried by the fire department are used more frequently than those in private industry. They are subject to harder use and to a greater variety of environmental conditions and vibrations **(Figure 1.26)**. To be protected, extinguishers should be carried within compartments rather than in an exposed location.

NFPA 1901, *Standard on Automotive Fire Apparatus,* requires that portable fire extinguishers be carried on apparatus. The extinguishers must be suitable for Class A, Class B, and Class C fires. Although individual fire extinguishers are specified by the apparatus purchaser, the minimum ratings for different types of extinguishers to be carried on fire apparatus are as follows:

- Two extinguishers rated for B:C fires. Dry chemical extinguishers must have at least an 80-B:C rating. Carbon dioxide extinguishers must have at least a 10-B:C rating.
- One 2½ gallon (10 L) or larger water extinguisher.

The exception to these requirements is that initial attack apparatus, due to space constrictions, are required to carry only one extinguisher rated for Class B and Class C rated fires. To date, NFPA does not require a Class K extinguisher on fire apparatus.

Inspecting, Maintaining, and Recharging Extinguishers

Proper servicing is essential to maintaining portable fire extinguisher readiness. The following sections highlight the procedures required for properly inspecting, maintaining, and recharging portable fire extinguishers.

Inspecting Extinguishers

In most occupancies, extinguishers are used so infrequently that there is a natural tendency to ignore them until a fire occurs. Therefore, regular inspections of extinguishers are very important to ensure their readiness. Unless they are inspected regularly, some of the following situations can impair extinguisher readiness:

• An extinguisher can be stolen or misplaced.

• An extinguisher can be damaged as a result of being struck by a vehicle such as a forklift truck **(Figure 1.27)**.

• An extinguisher may have lost its pressure for a variety of mechanical reasons.

• An extinguisher may have been used on a fire and then replaced on its mount without anyone notifying the proper authorities.

NOTE: Failure to notify authorities is more likely to occur if an employee is reluctant to report a fire to his or her employer.

Inspection and Maintenance

There is a distinction between inspection and maintenance. An inspection is a visual check to determine that an extinguisher is available and operable. Maintenance involves a more thorough examination and the performance of any service that is needed. A typical maintenance examination may involve actual operation of the extinguisher and fully dismantling it to inspect for worn parts. Most jurisdictions require some form of mandatory periodic maintenance.

Extinguisher inspection may seem like a trivial procedure, but it is not. An industrial complex may have hundreds of extinguishers; simply checking to ensure that they have not been stolen, vandalized, or are missing is an important part of plant protection. A single fire extinguisher that is missing or that is inoperable during an emergency may result in property loss or injury. Portable fire extinguisher inspections are usually performed by building personnel, but often may be performed by fire suppression or fire prevention officers **(Figure 1.28)**.

Figure 1.27 Damaged extinguishers need to be taken out of service.

Figure 1.28 Every fire extinguisher needs to be inspected for proper placement and operability.

NFPA 10 recommends that extinguisher inspections be performed monthly. Keeping accurate records of extinguisher inspections is a good business practice. The inspection date and the inspector's name or initials should be recorded on the extinguisher inspection tag; bar code readers may also be used to record the inspection (**Figure 1.29**). The inspection tag also provides the property owner with chronological data to verify compliance with codes and insurance requirements (**Figure 1.30**). In addition to the exterior service tag, an extinguisher may have a verification of service collar tag attached to it. The collar tag is made of polyethylene or aluminum and is tightened down against the opening of the extinguisher body. The purpose of this tag is to verify that the extinguisher was actually opened and discharged during maintenance. The collar tag will have maintenance data impressed into the plastic for the inspector's records. The collar tag must not be damaged.

NOTE: Consult NFPA 10, *Standard for Portable Fire Extinguishers*, for more information.

During an inspection, personnel should perform the following:

- Check that the extinguisher is in its proper location.

- Ensure that access to the extinguisher is not obstructed by boxes, clothing, storage items, or is otherwise inaccessible.

- Check the inspection tag to determine if maintenance is due.

- Examine the nozzle or horn for obstructions.

- Check lock pins or tamper seals to make sure that they are intact (**Figure 1.31**).

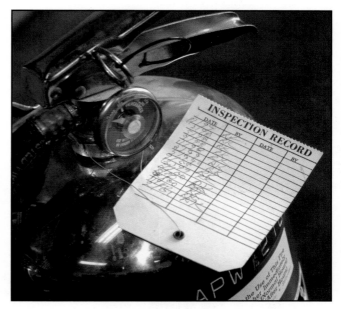

Figure 1.30 Chronological data helps to verify compliance with codes and insurance requirements.

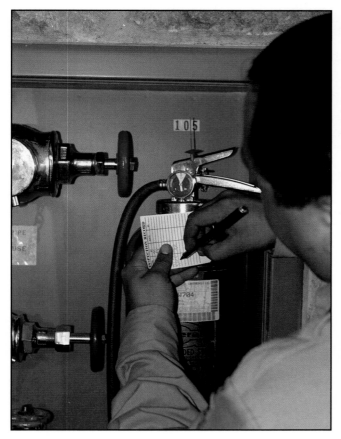

Figure 1.29 Keeping accurate records of extinguisher inspections is very important.

Figure 1.31 Lock pins need to be in place.

Figure 1.32 Make sure that the pressure gauge indicates that the extinguisher is at the correct operating pressure.

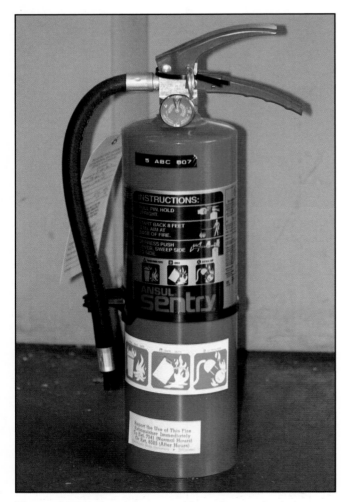

Figure 1.33 Extinguishers must have the required signage in place so that they can be readily identified.

- Check for signs of physical damage.
- Check that the extinguisher is full of agent.
- Check that the pressure gauge indicates proper operating pressure (**Figure 1.32**).
- Check collar tag for current information and/or damage.
- Check that required signage is in place (**Figure 1.33**).
- Check to see if the operating instructions on the extinguisher nameplate are legible.
- Check that the extinguisher is suitable for the hazard protected.

Maintaining Extinguishers

Extinguisher maintenance, as compared to simple inspection, should be performed whenever an inspection reveals the need for maintenance or the unit becomes due for interval maintenance required by state or local codes. For more information on maintenance intervals, see NFPA 10. The purpose of maintenance is to ensure that the extinguisher will operate safely. Because extinguisher maintenance requires specific technical knowledge and specialized tools and parts, it is usually performed by licensed service technicians employed by extinguisher distribution and service companies (**Figure 1.34**). In situations where a property owner has a large number of extinguishers, however, it may be more economical for in-house trained safety personnel to do the work. The job cannot be delegated to untrained personnel as a part-time duty.

Maintenance of portable extinguishers involves a thorough examination of the following basic elements:

- Mechanical parts
- Extinguishing agent
- Expelling means

It is not necessary to internally examine, on a yearly basis, stored-pressure extinguishers with pressure gauges or carbon dioxide extinguishers unless there is evidence of damage or leakage. When replacing parts, only parts obtained from the extinguisher manufacturer should be used.

NOTE: Wherever dry power-type fire extinguishers are mounted on vehicles, it is prudent to remove the extinguishers from vehicles periodically and shake or stir the contents. Vibration during vehicle operation can cause the contents to settle and pack, compromising the operation of the extinguisher when it is needed.

Recharging Extinguishers

Recharging is one of the most critical procedures in the maintenance of a fire extinguisher. Recharging is the replacement of the expellant and, if necessary, the agent **(Figure 1.35)**. A study conducted by the National Association of Fire Equipment Distributors (NAFED) showed that 6.7 percent of the extinguisher failures studied were the result of the extinguisher being improperly recharged. Recharging is not required on a periodic basis for every type of extinguisher. NFPA 10 provides for certain extinguishers to be maintained annually and others to be maintained every six years. For more information on fire extinguisher maintenance intervals, see NFPA 10.

When refilling an extinguisher, it is important that only those chemicals or materials specified by the manufacturer or those having an equivalent composition be used. For example, dry chemicals must be of proper particle size to flow properly. Although sodium bicarbonate is one of the dry chemicals used in extinguishers, the baking soda purchased from a supermarket is not suitable for use in an extinguisher. Different extinguishing agents cannot be mixed, and only the agent intended for a particular extinguisher can be used in that extinguisher. A dry powder agent, for example, cannot be used in a dry chemical extinguisher because of differences in flow characteristics.

Figure 1.34 Extinguisher maintenance needs to be performed by trained technicians.

Figure 1.35 It is critical that extinguishers be recharged properly.

Recharging some extinguishers requires not only refilling the unit with the proper agent but also pressurizing it as well. As with other aspects of extinguisher maintenance, pressurization must be performed using proper equipment and methods (**Figure 1.36**). NFPA 10 allows for the expelled agent to be reused, providing that it is the correct agent and is not contaminated. In the case of dry chemical or Halon extinguishers, pressurizing with compressed air is not satisfactory. This air typically contains moisture that can cause caking of dry chemicals and hydrolysis of Halon. Nitrogen should be used for pressurizing these units.

One obvious danger in pressurizing an extinguisher is applying too much pressure to the cylinder. It is important, therefore, to use a source of compressed air with a pressure not greater than 25 psi (175 kPa) above the operating pressure. Another potential danger is the inclusion of moisture in nonwater extinguishers. Moisture not only can result in the caking of dry chemical, but also can contribute to interior corrosion of the container and ultimate failure. Therefore, nonwater extinguishers must be thoroughly dried before being refilled. Finally, stored-pressure extinguishers must be subjected to a leak test after recharging (**Figure 1.37**).

Cartridge-operated extinguishers are the easiest type to refill. Once all the pressure has been expelled from the extinguisher, the extinguisher cap may be removed. The agent is poured out and stored in a sealed container while the rest of the maintenance is being performed. The agent is then poured back into the container to the mark provided on the inside of the tank (**Figure 1.38**). The lid is then securely replaced. The expended nitrogen cylinder is also replaced and the extinguisher is ready for service. **As with all fire extinguishers, only trained personnel should refill and recharge cartridge fire extinguishers.** Typically, this service will be provided by someone from an extinguisher service company, such as a state licensed technician.

Hydrostatic Testing of Portable Extinguishers

With the exception of pump-tank water fire extinguishers discussed earlier in this chapter, fire extinguishers are actually pressure vessels. They

Figure 1.36 Pressurization of extinguishers must be performed carefully.

Figure 1.37 A stored-pressure extinguisher being subjected to a leak test.

Figure 1.38 For cartridge-operated extinguishers, the agent can be poured back into the cylinder.

Figure 1.39 It is important to know the correct operating pressure for each type of fire extinguisher.

Table 1.6 Hydrostatic Test Intervals for Extinguishers	
Extinguisher Type	Test Interval (Years)
Stored-pressure water and/or antifreeze	5
Wetting agent	5
AFFF (aqueous film forming foam)	5
FFFP (film-forming fluoroprotein foam)	5
Dry chemical with stainless steel shells	5
Carbon dioxide	5
Wet chemical	5
Dry chemical, cartridge-or cylinder-operated, with mild steel shells	12
Halogenated agents	12
Dry powder, stored-pressure, cartridge- or Cylinder-operated, with mild steel shells	12

Note: Stored-pressure water extinguishers with fiberglass shells (pre-1976) are prohibited from hydrostatic testing due to manufacturer's recall.

Reprinted with permission from NFPA 10, *Standard for Portable Fire Extinguishers*, Copyright© 2002 National Fire Protection Association, Quincy, MA 02269. This reprinted material is not the complete and official position of the National Fire Protection Association on the referenced subject, which is represented only by the standard in its entirety.

either are maintained under a constant pressure or are pressurized when they are used. The pressure in extinguishers varies from 100 to 850 psi (700 kPa to 5 860 kPa), depending on the type (**Figure 1.39**). Physical damage and interior or exterior corrosion can cause extinguisher shell failure, which may result in severe injury or death. To ensure that an extinguisher is strong enough to withstand the pressures to which it is subjected, it must be periodically tested.

The method used to pressure test an extinguisher is known as the hydrostatic test. Hydrostatic testing consists of filling the cylinder with water and then applying the appropriate pressure by means of a pump. This method is used because it is safer than using a compressed gas. If the cylinder fails while pressurized with water, a violent rupture usually does not occur. Hydrostatic testing of extinguishers is performed at the intervals specified in **Table 1.6**, or whenever there is evidence of damage or corrosion. Hydrostatic testing should only be performed by trained and experienced personnel.

The hydrostatic test pressure is based on the extinguisher's service pressure and its factory test pressure. For carbon dioxide extinguishers and carbon dioxide or nitrogen cylinders used as an expellant source, the hydrostatic test pressure is 5/3 (167 percent) of the service pressure stamped on the vessel. The factory test pressure is the pressure at which the vessel was tested at the time of its manufacture. This pressure is shown on the extinguisher nameplate **(Figure 1.40)**. For Halon 1211 and stored pressure extinguishers, the hydrostatic test pressure is the factory test pressure.

An extinguisher should not be subjected to hydrostatic testing if it shows signs of physical defects such as damaged threads, corrosion, or welded repairs **(Figure 1.41)**. If an extinguisher vessel is of soldered or riveted brass or copper construction, or if calcium chloride has been used in a stainless steel shell and it has been burned in a fire, it should

not be hydrostatically tested. If an extinguisher vessel ever fails a hydrostatic test, it must be removed from service and destroyed.

As with inspections, record keeping of extinguisher maintenance and hydrostatic testing are important parts of a fire protection program. Extinguisher maintenance is recorded on a collar tag (plastic) that is secured to the neck of the extinguisher and/or a paper tag that is attached to the extinguisher handle or other hardware. The information on this tag/tags includes the month and year the maintenance was performed and the name of the person who performed the work. For noncompressed-gas extinguishers, hydrostatic tests are recorded on a metallic label attached to the shell of the extinguisher. For compressed-gas types and for expellant cartridges, the month and year of the test are stamped on the cylinder **(Figure 1.42)**. Along with the labels or tags attached to the extinguisher, a record-keeping system (cards, film, etc.) should be established for

Figure 1.40 The factory test pressure is shown on the extinguisher nameplate for carbon dioxide extinguishers.

Figure 1.41 An extinguisher should not be hydrostatically tested if it appears to be damaged or corroded.

Figure 1.42 The month and year of the test are stamped on compressed-gas cylinders and expellant cartridges.

Figure 1.43 Any cylinder that fails a hydrostatic test must be clearly marked as unusable and taken out of service immediately.

management purposes. If an extinguisher fails a hydrostatic test, it must be clearly marked as such and taken out of service (**Figure 1.43**).

Using Portable Extinguishers
General Techniques

To place an extinguisher into operation, the operator should follow four basic steps to extinguish the fire effectively and safely:

Step 1: Activate alarm system.

Step 2: Select a suitable extinguisher.

Step 3: Activate the extinguisher effectively.

Step 4: Apply the extinguishing agent properly.

Step 1: Activate the alarm system.

When a fire is discovered, the first action that should be taken by the person who discovers it should be to call the appropriate fire department. The second action that should be taken is to initiate

a local alarm that will alert other occupants of the situation (**Figure 1.44, p. 40**). Typically, a person who encounters a fire is not experienced or trained in fire fighting. This may not be the case in some large industrial organizations where employees receive such training, but it is true in the general population. Critical time can be lost if someone attempts to fight a fire and is unsuccessful.

Step 2: Select a suitable extinguisher.

Extinguishers should be chosen that minimize risk to life and property and are effective in extinguishing the fire. Both the picture-symbol and letter-symbol methods provide a quick means of matching an extinguisher to a class of fire (keeping in mind that the letter-symbol method is rapidly disappearing in favor of the universal picture-symbol method). In the excitement that may occur when an inexperienced person encounters even a small fire, however, the technical aspects of fire fighting can be overlooked. The best technique is to

Figure 1.44 An alarm should be sent before attempting to fight a fire of any size.

After help has been called, the extinguisher should be removed from its mounting or cabinet and brought to the area of the fire. The fire should be approached from the upwind side to minimize danger to the operator and to more efficiently extinguish the fire.

Step 3: Activate the extinguisher.

Because extinguishers must be activated quickly, extinguisher manufacturers strive to simplify extinguisher operation so that even untrained persons can use them effectively. Activation of an extinguisher usually involves only two or three steps. With a stored-pressure unit, activation consists of removing the safety pin and squeezing the valve handle. For cartridge-operated units, the nozzle is usually removed from its holder (**Figures 1.45 a and b**).

The extinguisher is then pressurized by depressing a lever that punctures the cartridge seal (**Figure 1.46**). The nozzle is directed at the fire and the valve lever is squeezed (**Figure 1.47**).

Wheeled fire extinguishing units are usually equipped with a longer hose, and the entire length of hose must be uncoiled from its rack (**Figure 1.48, p. 42**). These units are activated by depressing the quick-activation lever or by turning a hand wheel on the nitrogen cylinder (**Figures 1.49 a and b, p. 42**).

analyze the type of hazard being protected and to provide, in a conspicuous location, an extinguisher of the proper size and type. Thus, a Class K-rated fire extinguisher would be provided in the kitchen area of a restaurant while a pressurized water or multipurpose extinguisher would be provided in the coat-check room. This technique relieves an inexperienced person from having to choose the appropriate extinguisher during an emergency. **Table 1.7** will help to identify the types of extinguishers that are suitable for each class of fire.

	Table 1.7			
	Fire Extinguishers and Classes of Fire			
Class A	**Class A & B**	**Class B & C**	**Class D**	**Class K**
Pressurized Water	AFFF	Carbon Dioxide	Graphite	Potassium acetate, Potassium carbonate, Potassium citrate
AFFF	Halon 1211	Halon 1211 Halon 1301	Sodium carbonate, Sodium chloride	Extinguishers with Class B rating OK but not recommended
Halon 1211	Multipurpose dry chemical	Multipurpose dry chemical	Ternary eutetic chloride (TEC)	
Multipurpose dry chemical		Potassium bicarbonate, Potassium chloride, Sodium bicarbonate	Trimethoxyboroxine	Antifreeze

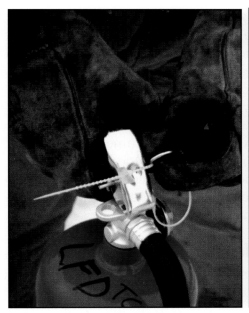

Figures 1.45 a and b Activating an extinguisher. For a stored pressure unit, remove the safety pin and squeeze the valve handle. For a cartridge-operated unit, remove the nozzle from its holder.

Figure 1.46 Depressing the lever punctures the cartridge seal and pressurizes the unit.

Figure 1.47 Direct the nozzle at the fire and depress the valve lever.

Figure 1.48 When using a wheeled unit, uncoil the entire length of hose.

Step 4: Apply the extinguishing agent properly.

The effective use of fire extinguishers depends on the technique of the operator. Many incipient fires can be extinguished by a brief discharge from the extinguisher. However, the total discharge duration for many extinguishers is less than 30 seconds. An excited novice operator can easily use an extinguisher in an inefficient manner. If the agent is misdirected or discharged at too great a distance, extinguishment will not be completed. If the fire is not completely extinguished, a more dangerous situation can develop.

P.A.S.S. Method

The P.A.S.S. method of operating a portable fire extinguisher is an easy way to remember the correct order in which to activate and operate an extinguisher (**Figure 1.50**). It is useful as a teaching tool for either firefighters or the general public. In a fire emergency, every second is of great importance; therefore, everyone should be acquainted with the general instructions applicable to most portable fire extinguishers. The acronym P.A.S.S. is used as follows:

P — Pull the pin at the top of the extinguisher. Break the plastic or thin wire inspection band as the pin is pulled.

A — Aim the nozzle or outlet toward the fire. Some hose assemblies are clipped to the extinguisher body. Release the hose and point.

Figures 1.49 a and b Wheeled units are activated by depressing the quick-activation lever or by turning a hand wheel on the nitrogen cylinder. *Photo b courtesy of Tyco Safety Products.*

Figure 1.50 The four-step method of extinguisher usage: Pull, Aim, Squeeze, and Sweep.

S — Squeeze the handle above the carrying handle to discharge the agent. The handle can be released to stop the discharge at any time. Before approaching the fire, try a very short test burst to ensure proper operation.

S — Sweep the nozzle back and forth at the **base of the flames** to disperse the extinguishing agent. After the fire is out, watch for remaining smoldering hot spots or possible reflash of flammable liquids. **Make sure that the fire is out.**

Attacking Class A Fires

When using a water-based extinguisher, the stream must be aimed at the seat of the fire to maximize the cooling effect of the water on the fuel. Initially, the extinguisher should be used at a distance of 10 to 30 feet (3 m to 10 m) from the fire (**Figure 1.51**).

CAUTION!

Attacking a fire from a distance of more than 30 feet (10 m) with a pressurized water fire extinguisher will be the least effective method. The distance of 30 feet (10 m) should be considered a starting point for an inexperienced operator.

When the flames are knocked down, the operator should move closer to wet down any remaining smoldering materials (**Figure 1.52, p. 44**). Fires involving compacted fuels or other deep-seated burning materials must be thoroughly soaked and should be pulled apart to reach the remaining fire. The extinguisher can be used intermittently to facilitate the soaking. If possible, the material should be moved outside to complete the overhaul process. A thumb or finger can be placed partially over the nozzle orifice to break up the stream into a spray pattern if desired.

AFFF extinguishers are effective against Class A fires and are used in a manner similar to water extinguishers. In addition to cooling, the AFFF has

Figure 1.51 When using a water-based extinguisher, it is important to move in close enough that the agent can be aimed at the seat of the fire: 10 to 30 feet (3 m to 10 m).

Figure 1.52 When the flames have been knocked down, move in closer to extinguish any smoldering materials.

a low surface tension, which enhances its ability to penetrate into fuels, especially tightly packed fuels.

When using a multipurpose dry chemical extinguishing agent (monoammonium phosphate), the fire should be attacked at its base, sweeping the nozzle from side to side. Because the multipurpose agent forms a coating that excludes oxygen on the fuel, it is important that the dry chemical agent thoroughly coat all fuel surfaces. The dry chemical has little cooling effect and deep-seated fires may prove difficult to extinguish.

Multipurpose dry chemical extinguishers are also useful on lightweight materials when other types of extinguishers might scatter the material and spread the fire. For this type of attack, the extinguisher is discharged approximately 10 feet (3 m) from and 3 feet (1 m) above the fire. This agent "cloud" forms a covering over the fire and the surrounding area **(Figure 1.53)**.

Halon 1211 portable fire extinguishers have limited application with Class A fires. Extinguishers having a 9-pound (4 kg) or greater capacity have a Class A rating. Halon 1211 is most effective in suppressing the flaming portion of combustion involving Class A fuels. It is less effective against the

Figure 1.53 Multipurpose dry chemicals are discharged about 3 feet (1 m) above the fire to cover the fire.

deep-seated portion of the combustion, requiring a higher concentration of agent for complete control. When attacking a Class A fire with a Halon 1211 fire extinguisher, the operator should avoid breathing the vapors produced by the thermal decomposition of the agent.

Attacking Class B Fires

Both regular and multipurpose dry chemical extinguishers are used to extinguish fires involving flammable or combustible liquids and gases. The extinguishing agent should initially be discharged from a distance of approximately 10 feet (3.1 m). If the attack is started at a closer range, the velocity of the dry chemical discharge may cause the fuel to splash, thus spreading the fire. In general, a flammable liquid fire should be attacked by sweeping the leading edge of the fire with the agent and moving forward to continue applying the agent (**Figure 1.54**). This action interrupts the chemical chain reaction and reduces the radiant heat. When fighting flammable liquid fires, it is possible for the fire to flash back across the surface of the liquid. This situation occurs because the extinguishing action of the dry chemical – the interruption of the chemical reaction — is not cumulative as it is with foam. Therefore, an operator must stay alert and be prepared for immediate retreat if a flashback occurs.

If it is necessary to attempt a second attack on the fire, it must be done with a second full extinguisher. Flammable liquid fires have the characteristic of being particularly fast spreading. If any doubt exists about the ability to control the fire, personnel should leave the area and await the arrival of the fire department.

Carbon dioxide extinguishers are also effective on flammable liquid fires. Because it is a gas, carbon dioxide cannot be projected very far out of the nozzle; therefore, it must be applied at a closer range than other agents (**Figure 1.55**). The agent should be applied by sweeping it across the surface of the burning liquid, overlapping it. Carbon dioxide has a cooling as well as a smothering effect. Consequently, discharge should be continued after initial extinguishment to cool the fuel and prevent flashback.

AFFF is effective on Class B fires involving hydrocarbons such as fuel oils, gasoline, and kerosene. AFFF is not effective, however, on flammable liquids such as acetone, alcohols, ethers, and lacquer thinner, because they can break down the foam. It is also not effective on pressurized liquids and gases.

Because AFFF extinguishes by establishing an air-excluding barrier, the application technique must enhance the formation of a surface film.

Figure 1.54 For a flammable liquid fire, sweep the leading edge of the fire and move forward to continue applying the agent. *Courtesy of Tyco Safety Products.*

Figure 1.55 Carbon dioxide cannot be projected very far from an extinguisher, so it must be applied close to the fire. *Courtesy of Tyco Safety Products.*

Figure 1.56 Two ways in which AFFF can be applied.

AFFF should not be discharged directly into the liquid surface because it will penetrate the burning liquid and cause splashing. On fires of depth, the foam should be deflected off the sides or back of the enclosing tank so that the agent will flow down onto the liquid **(Figure 1.56)**. On spill fires, the AFFF can be directed onto the surface just in front of the fire to spray over the fire.

Attacking Class C Fires

When attacking fires involving energized electrical equipment, the primary consideration is to ensure that the agent is dielectric or electrically nonconductive so that the operator is not injured. If possible, the equipment should be de-energized before initiating the attack. Whether de-energized or not, fire involving electrical equipment can be fought effectively with Halon, carbon dioxide, dry chemical, or even a "water mist" extinguisher. Originally designed to take the place of Halon-type fire extinguishers, the distilled water fire extinguisher has proven to be effective in extinguishing Class C fires.

If a dry chemical extinguisher is used, there will be a need for substantial cleanup. Dry chemical agents may also damage sensitive electronic equipment that is in the area but is not involved in the fire, so their use should be avoided around such equipment if possible. Carbon dioxide or water mist extinguishers are the best choices for use on sensitive electrical equipment. Both agents are nonconductive, noncorrosive, and leave no residue. Carbon dioxide should be applied at close range for a quick knockdown. Water mist stored-pressure fire extinguishers can be used from a somewhat greater distance (up to 12 feet [4 m]). The application wand on a water mist-type extinguisher is designed for operator safety, as is the agent (distilled water) itself. Halon fire extinguishers are also effective on Class C fires, but are destructive to the environment and are toxic to the operator. As mentioned earlier in this chapter, Halon fire extinguishers are still seen in the field but once used are considered nonrechargeable unless certain essential circumstances are shown.

Attacking Class D Fires

Various dry powder agents will extinguish fires involving magnesium, sodium, and potassium alloys. Each agent has its limitations, however, and these must be known before attacking a Class D fire. Even though a particular agent may work well in combating lithium fires, for example, the same agent may not be effective on magnesium fires. Most dry powder agents extinguish combustible metals by caking and adhering to the material, which excludes the necessary air required for combustion. The burning material should be covered with a 2-inch (50 mm) layer of dry powder agent.

Figure 1.57 Dry powder agent being moved onto a piece of burning metal.

Dry powder agents should be applied with a minimum of disturbance to the burning material; most are applied with either an extinguisher or a shovel (**Figure 1.57**).

Attacking Class K Fires

Class K (cooking media) fires are particularly difficult to extinguish because of their tendency to reignite after the fire has been extinguished. It has been discovered that although a Class K fire may have been extinguished properly with a dry chemical, the fuel changes chemically and reaches auto-reignition at a lower temperature. For this reason, only an extinguisher with a Class K rating is recommended for use on this type of fire. A Class A:B:C or Class B:C rated fire extinguisher may indeed work effectively, but the operator should be on alert for a possible auto-reignition and a second or third attack on the fire.

The initial attack on a Class K fire with an appropriately marked Class K extinguisher is similar to other types of attack. Begin application from a distance of 10 – 12 feet (3.1 to 4 m) away from the burning material, hold the application wand at the edge of the flames, and coat the surface of the material with a side-to-side sweep. Continue to apply agent until the fire extinguisher is completely empty. Extinguishment takes place through removal of oxygen to the fire and partially through cooling the fuel. It is the innate cooling quality of the agent, along with its ability to form soapy foam (saponification), that prevents the fuel from reaching a lower reignition temperature.

Summary

Portable extinguishers have long been recognized and used as a first line of defense against incipient fires. When extinguishers are correctly selected and placed for the hazards they are to control, inspected and maintained regularly, and used with proper training, they have proven to save lives and property.

Not all hazards are alike and therefore not all portable extinguishers are alike. In addition, extinguishers are useless if they are unavailable, are not designed for a particular hazard, or do not work because of poor maintenance or damage. Fire fighting personnel have a large role to play in selecting the correct extinguisher and in seeing that it is placed so that it is accessible during an emergency. These skills require knowledge of the operating principles of various extinguishers, the advantages and disadvantages of different extinguishing agents, and familiarity with inspection and maintenance procedures. Personnel must also be on the alert for obsolete or damaged extinguishers so that they can be removed from service before they cause injury when someone attempts to use them. Finally, individuals in a workplace need to be trained in the basics of extinguisher operation so that they may be able to control a fire and evacuate the premises until additional help arrives.

Fire Detection and Signaling Systems

1. Describe the basic components of a modern fire detection and signaling system.

2. List the power supply systems that are required on fire detection and signaling systems.

3. List the types of initiating devices that are used in fire detection and signaling systems.

4. List the common types of signaling systems and describe their operation.

5. Describe the factors that determine which type of fire detection and signaling system should be installed in an occupancy.

6. Describe the operation of an emergency voice/alarm communication system.

7. Describe the function and operation of manual alarm-initiating devices.

8. Identify different automatic alarm-initiating devices.

9. Describe the operating principles of heat, flame, fire gas, and combination fire detectors.

10. Describe the steps necessary to complete a fire detection and signaling system service test and periodic inspection.

11. List the types of records and reports that should be included in an agency's document management system.

FESHE Objectives

FESHE Objectives
Fire and Emergency Services Higher Education (FESHE) Objectives:
Fire Science Curriculum: Fire Protection Systems

- Classify detection, alarm, supervisory devices, heat, flame, smoke control devices and hardware.

Chapter 2
Fire Detection and Signaling Systems

The early detection of a fire and the signaling of an appropriate alarm remain the most significant factors in preventing large losses from occurring. History has proven that delays in fire detection and alarm transmission lead to increased injuries, deaths, and property losses **(Figure 2.1)**. Modern fire detection and signaling systems, if properly installed and maintained, are a reliable method of reducing the risk of a large-loss incident.

This chapter provides information on the basic components of fire detection and alarm systems. Covered in more detail are the various types of systems available and the devices that actuate the alarm signal. The last portion of the chapter highlights the procedures that fire brigade and fire department or other personnel should use for inspecting, testing, and record keeping of fire detection and alarm systems. The importance of accurate record-keeping activities is also discussed.

Basic System Components

The modern detection and signaling system can be a very elementary system or can feature very advanced detection and signaling equipment **(Figure 2.2, p. 52)**. Such a system can be designed by a fire protection engineer or designed and installed by technical personnel employed by a fire alarm system company. The design, installation, and approval of a fire detection and signaling system may also require approval by various code authorities before it is accepted as an in-service unit. A nationally recognized testing laboratory, such as Underwriters Laboratories Inc. (UL) or FM Global, should test the components of a system to ensure operational reliability **(Figure 2.3, p. 52)**. Testing reports

Figure 2.1 Early detection of fire can help prevent deaths, injuries, and property losses. *Courtesy of District Chief Chris E. Mickal.*

may address either an entire system or individual components that may be used in interchangeable applications. The installation of the system should conform to the applicable provisions of NFPA 70, *National Electrical Code®*, NFPA 72, *National Fire Alarm Code®*, and local codes and ordinances. Some standards are addressed later in this chapter within the discussions of the various types of

Figure 2.2 Modern fire alarm control panels are found on many commercial properties. *Courtesy of Tom Jenkins.*

Figure 2.3 Any component of an alarm or signaling system should be tested by a recognized authority.

systems. The following sections highlight each of the basic components that may be found in various types of fire detection and alarm systems.

NOTE: Some fire alarm control units are designed for both security and fire protection. In these types of systems, fire protection is engineered into the system to assume the highest priority.

Fire Alarm Control Unit

The fire alarm control unit is essentially the "brain" of the system **(Figures 2.4 a and b).** It is responsible for processing alarm signals from actuating devices and transmitting them to the local or other alerting system, such as an alarm company with telecommunicators on duty. In actual installations, the fire alarm control unit is often referred to as the *alarm panel.* All the controls for the system are located in the fire alarm control unit.

Power Supply

Electric power sources provided for operation of fire alarm systems must, of course, be adequate for the capacity of the system design. Discussed below are requirements of NFPA 72, *National Fire Alarm Code®,* for the primary power supply and the secondary power supply.

Primary Power Supply

The primary electrical power supply usually consists of the building's main connection to the local public electric utility. An alternative power supply is an engine-driven generator that provides electrical power. If such a generator is used, either a trained operator must be on duty 24 hours a day or the system must contain multiple engine-driven generators. One of these generators must always be set for automatic starting. Either power supply must be supervised and must signal an alarm if the power supply is interrupted.

Secondary Power Supply

A secondary power supply must be provided for the detection and signaling system. This is done to ensure that the system will be operational even if the main power supply fails. The secondary system must be able to make the detection and signaling system fully operational within 30 seconds of the

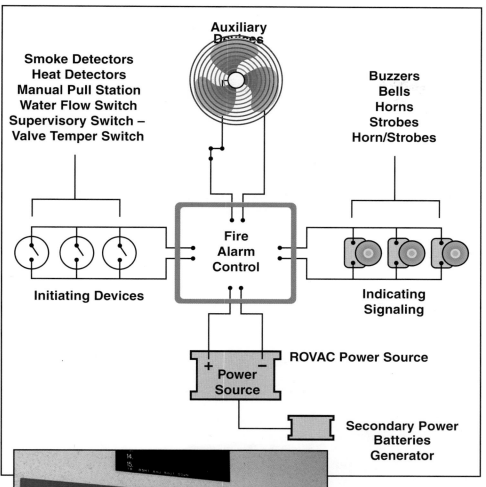

Smoke Detectors
Heat Detectors
Manual Pull Station
Water Flow Switch
Supervisory Switch –
Valve Temper Switch

Auxiliary
Devices

Buzzers
Bells
Horns
Strobes
Horn/Strobes

Fire
Alarm
Control

Initiating Devices

Indicating
Signaling

ROVAC Power Source

+   –
Power
Source

Secondary Power
Batteries
Generator

Figures 2.4 a and b The fire
alarm control unit is the brain
center of the alarm and detection
system.

main power supply's failure. **Table 2.1, p. 54,** shows
the capabilities that the secondary power source
must have. The secondary power source must consist of one of the following:

- Storage battery and charger **(Figure 2.5, p. 54)**

  NOTE: This cannot be a dry-cell battery.

- Engine-driven generator and a four-hour-capac-
  ity storage battery **(Figure 2.6, p. 54)**.

- Multiple engine-driven generators, of which one
  must always be set for automatic starting

Trouble Signal Power Supply

When a trouble signal is initiated by a fire alarm
system to indicate an electrical failure in a circuit
or component, the trouble signal indicator must
have a source of power. This source of power may
be the secondary power supply. In addition, it can
be a totally independent power supply, as long as it
does not entail the use of dry-cell batteries.

Table 2.1
Secondary Power Supply Requirements

System Type	Maximum Normal Load	Alarm Load
Local Systems	24 hours	5 minutes
Auxiliary systems	60 hours	5 minutes
Remote station systems	60 hours	5 minutes
Proprietary systems	24 hours of normal traffic	
Emergency voice/alarm communication systems	24 hours	2 hours

Figure 2.5 The alarm system must contain a storage battery and charger.

Figure 2.6 An engine-driven generator may be a component of the secondary power source.

Initiating Devices

Initiating devices are the manual and automatic devices that are activated or that sense the presence of fire and then send a signal to the system control unit. The initiating device may be connected to the fire alarm control unit by a hard-wire system or it may be radio-controlled over a special frequency. Initiating devices include manual pull stations, heat detectors, smoke detectors, flame detectors, waterflow devices, tamper switches, and combination detectors. These devices are covered in more detail later in this chapter.

Notification Appliances

Once an initiating system activates and sends a signal to the fire alarm control unit, the signal is processed by the control unit and appropriate action is taken. This action may include the sounding and lighting of local alarms and the transmission of an emergency signal to a proprietary system, central station service, or a fire department telecommunications center. Local notification devices include bells, buzzers, horns, recorded voice messages, strobe lights, speakers, and other warning appliances **(Figures 2.7 a - d)**. Depending on the design of the system, the local alarm may sound only in the area of the activated detector or it may sound in the entire facility.

Auxiliary Services

Some occupancies have special requirements in the event of a possible fire condition. In these cases, the fire detection and alarm system can be designed to perform the following special functions:

- Shut down or exhaust the heating, ventilation, and air-conditioning (HVAC) system for smoke control.

- Close smoke and/or fire doors and dampers **(Figure 2.8)**.

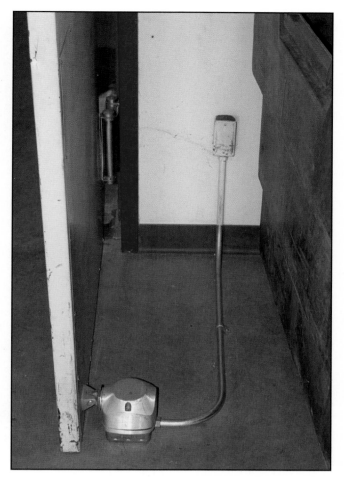

Figure 2.8 Automatic door closing devices can be part of the detection and alarm system.

Figures 2.7 a – d Notification devices include bells, horns, strobe lights, and speakers.

Figure 2.9 A deluge system may be necessary in certain occupancies.

- Pressurize stairwell(s) for evacuation purposes.
- Override control of elevators and prevent them from opening on the fire floor.
- Automatically return the elevator to a designated evacuation floor.
- Operate heat and smoke vents.
- Activate special fire extinguishing systems, such as preaction and deluge sprinkler systems, or a variety of fixed extinguishing agents (**Figure 2.9**).
- Monitor certain aspects of fire pump, pump driver, and generator.

Types of Signaling Systems

The purpose of a protective signaling system is to limit fire losses involving life and property. Signaling systems vary from the very simple to the very complex (**Figures 2.10 a and b**). A simple system may only sound a local evacuation alarm whereas a complex system may sound a local alarm, activate building services, and notify fire and security agencies to respond. The type of system installed in a given occupancy depends on any or all of the following factors:

- Level of life safety hazard
- Structural features of the building
- Level of hazard presented by the contents of the building
- Availability of fire suppression resources
- State and local code requirements

The following sections examine each of the common types of signaling systems that firefighters are likely to encounter. It is important that personnel be able to recognize each type of system and understand how it operates. This is particularly important when performing pre-incident planning at facilities with complicated systems. The major types of systems include the following:

- Protective Primises (Local)
- Auxiliary

Figures 2.10 a and b
Alarm systems may be simple or very complex.

- Remote Receiving
- Proprietary
- Central station

The requirements for all fire alarm and protective signaling systems are contained in NFPA 72, *National Fire Alarm Code®*.

Protected Premises (Local)

A protected premises (local) alarm system is designed to transmit both a visible and an audible alarm only on the immediate premises. There are no provisions for off-site reporting. The alarm's purpose is to alert the building's occupants and to ensure their life safety. The local system can be activated by manual means (pull station), or by automatic devices such as smoke detectors. A local system may also be capable of supervising itself to ensure that no service interruptions go unnoticed. Local systems can be designed to activate the auxiliary services described at the end of the previous section. There are three basic types of local alarm systems:

- Noncoded
- Master control unit
- Zoned/annunciated

Any of these types of local alarm systems may be equipped with a presignal alarm. Presignal alarms are employed in such locations as hospitals, where greater assistance is needed to help occupants evacuate in a safe and orderly manner. The system initially responds with a presignal that alerts emergency personnel before the general occupancy is notified. This presignal is usually a discreet signal that is recognizable only by personnel who are familiar with the system. The presignal may be a recorded message over an intercom, a soft alarm signal, or a pager notification. The presignal provides emergency personnel with an opportunity to assist the general occupancy in evacuation. Depending on the policies of the occupancy and local code requirements, the emergency personnel may elect to handle the incident without sounding a general alarm. They may sound the general alarm after investigating the problem, or the general alarm will sound automatically after a certain amount of time has passed and the fire alarm control unit has not been reset.

Noncoded Local Alarm

The noncoded system is the simplest type of local alarm. When an alarm-initiating device sends a signal to the system control unit, all the alarm-indicating devices operate simultaneously **(Figure 2.11, p. 58)**. The indicating devices usually operate continuously until the fire alarm control unit is reset. The fire alarm control unit is not capable of determining which indicating device triggered the alarm; therefore, building and fire department personnel must walk around the entire facility and visually check to see which device was activated. These systems are only practical in small occupancies with a limited number of rooms.

Master Control Unit

A master control unit serves the premises as a local control unit and receives input from other fire alarm control units **(Figures 2.12 a and b, p. 59)**. This system is used in occupancies that use the alarm signals for other purposes, such as a school

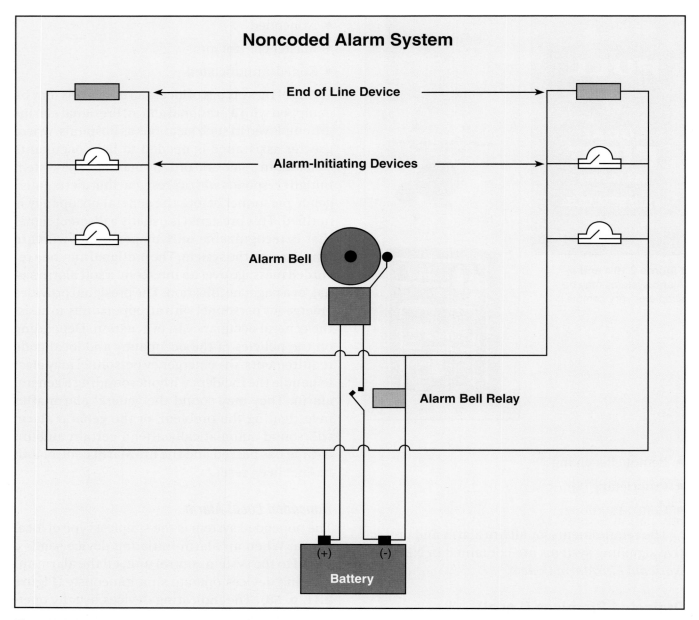

Figure 2.11 A noncoded system cannot differentiate which alarm-initiating device sent the alarm or produce anything other than a continuous signal.

that uses the same bells for class changes and fire alarms. A master control unit enables the fire alarm to have a sound that is distinct from the class bells, thus eliminating confusion as to which type of alarm is sounding.

NOTE: Newer codes do not allow co-mingling of fire alarm bells with other school bells, but older systems are still in existence. Often, a series of five or six short rings can sound in the same time that a normal long continuous ring would take.

On simpler systems, it could be a relay that opens and closes, causing an intermittent alarm signal. On a more complicated system, the alarm may be a series of short and long signals. Whichever the case, it must be remembered that with master control units, all fire alarms will sound the same alarm regardless of the location of the initiating device that triggered the alarm.

Figures 2.12 a and b A master control system is used in occupancies that need to use an alarm system for several purposes.

Zoned/Annunciated Alarm

The primary purpose of fire alarm system annunciation is to enable responding personnel to identify the location of a fire quickly and accurately in order to help ensure the safety of occupants. Where required, the location of an operating notification appliance should be visibly indicated by building, floor, fire zone, or other area by means of an annunciator or printout. In this type of system, the alarm-initiating devices in common areas are arranged in circuits or zones, and are wired into the fire alarm control unit as such (Figure 2.13). Each zone has its own indicator lamp on the fire alarm control unit. When an initiating device in a particular zone is triggered, the alarm sounds and the corresponding lamp on the control unit lights up. This signal gives responders a better idea of where the problem area is located. The audible alarm is a continuous alarm, just as with the non-coded system.

An *annunciator panel* may be located some distance from the fire alarm control unit, often in a location designated by the fire department.

Figure 2.13 Zoned systems are designed to help emergency responders quickly identify the location of a fire.

Such an installation may be found at the driveway approach to a large retirement complex, for example. This type of annunciator panel usually has a map of the complex coordinated to the zone indicator lamp. Arriving firefighters use the information provided on the annunciator panel to take the appropriate routes to the exact building involved and radio to incoming units their size-up information. Another type of annunciator panel may be found in the lobby area of the complex. It will have a graphic display of the involved area.

A zone indicating system may also be equipped with its own indicator lamp on the fire alarm control unit, a signal-coding device that is placed into the circuit. This device causes the indicating devices to sound in a specific and unique pattern for each zone. This pattern enables employees or fire department personnel to determine the problem zone simply by listening to the pattern of the alarms.

Usually, the pattern is a series of short rings, a brief pause, then a second series of short rings, followed by a long pause. Then the cycle will repeat. Most systems are designed to sound the alarm for the first zone that comes in and then disregard all subsequent zones that initiate alarm signals. Zoned and annunciated features are often found in conjunction with proprietary fire alarm systems, which are discussed later in this section.

Zone coded local alarm signals in local alarm systems are sometimes confused with a selective coded system that uses telegraphic fire department alarm boxes (**Figure 2.14**). However, the two are not the same. Zone coded systems produce a coded audible and/or visible alarm as a result of electronic circuitry located in the alarm panel. The alarm-initiating device only transmits the presence of an open circuit. The location of the activated alarm-initiating device in the system and the subsequent alarm panel circuit determine the code pattern that will be sent. However, a selective coded system transmits a coded signal to the alarm receiving station. This is accomplished through a coder device that is incorporated directly into the alarm-initiating device.

There are three basic devices that provide coded signals. The oldest method is the eccentric wheel (**Figure 2.15**). The wheel is mounted on a drive motor that is either spring or electronically powered; a set of flexible contacts rests on the wheel. The motion of the wheel results in the contacts opening and closing, causing a coded signal to be sent.

The second method uses electronic relay circuits. The relays are interconnected so that a current flowing through the circuit causes the relays to open and close in a predetermined pattern. Simple forms of these systems are similar to the turn-signal control on an automobile.

Newer systems use computer-aided devices that electronically process the information sent from the initiating device. The signal is then sent to the notification appliances and the receiving station in a distinctive pattern.

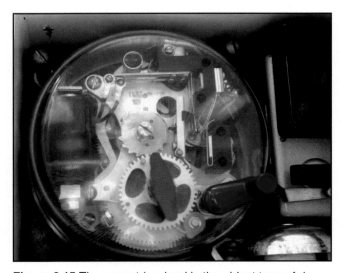

Figure 2.14 A selective coded system transmits a coded signal to the alarm receiving station.

Figure 2.15 The eccentric wheel is the oldest type of device that can provide coded signals.

Auxiliary Fire Alarm System

An auxiliary fire alarm system is defined as a facility that receives signals and at which personnel are in attendance at all times to respond to these signals. An example of an auxiliary fire alarm system is a municipal or county 9-1-1 dispatching facility or a security officer at a central control panel. An auxiliary fire alarm system is attached directly to a supervising station that may be of the hard-wired or radio-box type (**Figure 2.16**). When an alarm is activated in the protected occupancy, it is transmitted directly over the same communications lines that transmit alarms from street fire alarm boxes. This results in an alarm signal being received directly at the supervising station. The alarm can be initiated by manual pull stations, automatic fire detection devices, or waterflow indicating devices. Each community has its own requirements and policies for the use of these systems, although some do not allow them at all. Newer technologies will make the use of the auxiliary fire alarm system more rare. There are three basic types of these systems still found today:

- Local energy system
- Shunt system
- Parallel telephone system

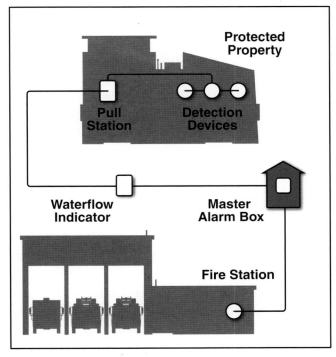

Figure 2.16 Supervising station alarm systems are becoming less common in today's society.

Local Energy System

A local energy system has its own source of power and is not dependent on the supply source that powers the entire municipal fire alarm system. Initiating devices can be activated even when the power supply to the municipal system is interrupted. However, this may result in the alarm only being sounded locally and not being transmitted to the fire department dispatch center. The ability to transmit alarms during power interruptions is dependent on the design of the municipal system.

Shunt System

A shunt system is electrically connected to an integral part of the municipal fire alarm system and is dependent on the municipal system's source of electric power. Should a power failure occur in this type of system, an alarm indication (false alarm) will be sent to the fire department receiving point. NFPA 72 allows only manual pull stations and waterflow detection devices to be used on shunt systems. Fire detection devices are not permitted on a shunt system.

Parallel Telephone System

A parallel telephone system is used on the type of municipal fire alarm system where each alarm box is connected to the fire department dispatch center by an individual circuit. The occupancy fire alarm system is also connected to the fire department dispatch center by an individual circuit.

Remote Receiving System

A remote receiving system is common in localities that are not served by central station systems. Instead of being hooked to the dispatch center through a municipal fire alarm box system, the remote system is connected by another means, usually a leased telephone line. Where permitted, a radio signal over a dedicated fire department frequency may also be used.

A remote receiving system is similar to an auxiliary system in that it is connected directly to the fire department dispatch center (**Figure 2.17, p. 62**). Even though a remote system may be located at an unmanned facility, it should have local alarm capability. The system will have the ability to transmit a trouble signal to the remote site when the system becomes impaired.

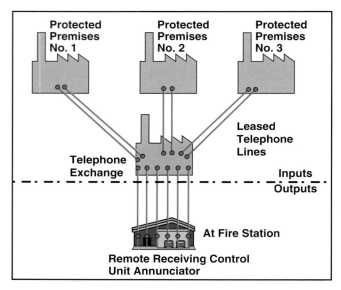

Figure 2.17 A remote receiver connects directly to the fire department dispatch center.

Depending on local preferences, the fire department may allow other organizations to monitor the remote system. In many small communities, the system is monitored by the local police agency at its dispatch center. This arrangement is particularly common in communities that have volunteer fire departments whose stations are not continuously staffed. In these cases, it is important that police dispatch personnel are aware of the importance of these alarm signals and are trained in the actions that need to be taken upon receiving them.

Proprietary System

A proprietary system is used to protect large commercial and industrial buildings, high-rises, and groups of commonly owned facilities in a single location, such as a college campus or industrial complex. Each building or area will have its own system that is wired into a common receiving point somewhere on the facility. The receiving point must be in a separate structure or in a part of the structure that is remote from any hazardous operations. The receiving station is continuously staffed by representatives of the occupant who are trained in the system's operation and which actions to take when an alarm is received (**Figure 2.18**). The operator should be able to automatically summon a fire department response through use of system controls, or may do so manually by using the telephone.

Figure 2.18 A proprietary system is staffed by trained personnel. *Courtesy of Paul Ramirez.*

Modern proprietary systems can be very complex systems that have a wide range of capabilities, including the following:

- Coded-alarm and trouble-signal indications
- Building utility controls
- Elevator controls
- Fire and smoke damper controls

Many proprietary systems and receiving points are used to monitor security functions in addition to fire and life safety functions.

Central Station System

A central station system is similar to a proprietary system. The primary difference is that instead of having the receiving point for alarms on the protected premises and monitored by the occupant's representative, the receiving point is at an outside, contracted service point called a central station (**Figure 2.19**). Typically, the central station is a company that sells its services to many individual customers. When an alarm is activated at a particular client's location, central station employees receive that information and initiate an emergency response. This involves calling the fire department and representatives of the occupancy. The alarm systems at the protected property and the central station are most commonly connected by dedicated telephone lines. All central station systems should meet the requirements set forth in NFPA 72, *The National Fire Alarm Code®.*

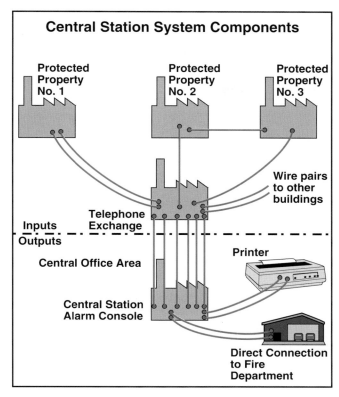

Central Station System Components

Protected Property No. 1

Protected Property No. 2

Protected Property No. 3

Wire pairs to other buildings

Telephone Exchange

Inputs
Outputs

Central Office Area

Printer

Central Station Alarm Console

Direct Connection to Fire Department

Figure 2.19 Components of a central station system.

Emergency Voice/Alarm Communications System

An emergency voice/alarm communications system is not a separate type of signaling system. It is a supplementary system that may be placed in a facility in addition to one of the other types of systems previously discussed. The purpose of this system is to provide the ability to communicate detailed information to occupants and fire fighting personnel who are in the facility. This system may be a stand-alone system, or it may be integrated directly into the overall fire detection and signaling system. Occupant notification must be part of the system. There are two basic types of emergency voice/alarm communications systems: one-way and two-way. One-way systems are most commonly used to warn building occupants that action is needed and to tell them what actions need to be taken (**Figure 2.20**). People can be directed to move to a different portion of the building, they can be ordered to leave the building, or if in unaffected areas, they may be told to stay where they are. Two-way systems allow people at other locations in the building to interact with the person at the main control station. This is accomplished by us-

Figure 2.20 One-way voice alarm systems are used to alert and instruct building occupants during an emergency.

Figure 2.21 An emergency phone located in the stairwell allows communication with the main control station.

ing either intercom controls or special telephones. This type of system is most helpful to fire fighting personnel who are operating in the building, particularly in buildings that interfere with portable radio transmissions. The emergency phones are located in the stairwells or near elevators, and they allow fire fighting crews to communicate with the incident commander at the main control station (**Figure 2.21**).

Manual Alarm-Initiating Devices

Manual alarm-initiating devices, commonly called manual pull stations or pull boxes, are placed in structures to allow occupants to manually initiate the fire signaling system. Manual pull stations may be connected to systems that sound local alarms, off-premise alarm signals, or both.

General Requirements

Although manual pull stations come in a variety of shapes and sizes, they are usually red in color with white lettering that specifies what they are and how they are to be used **(Figure 2.22)**. The fire alarm manual pull station should be used for fire signaling purposes only unless it is designed for a multipurpose use, such as to communicate with a guard station or to activate a fixed suppression system. According to NFPA 72, the pull station should be mounted on walls or columns so that the operable part is no less than 3½ feet (1 m) and no more than 4½ feet (1.5 m) above the floor. The manual pull station should be positioned so that it is in plain sight and unobstructed. Multistory facilities should have at least one pull station on each floor. In all cases, travel distances to the manual pull station should not exceed 200 feet (60 m). NFPA 72 also requires that pull stations be placed within 5 feet (1.5 m) of every exit so that facility occupants can activate an alarm while they are exiting the facility **(Figure 2.23)**. Most codes do not require manual alarm-initiating devices in structures that are fully sprinklered with a system that sounds a local alarm when water begins to flow.

Older-type manual pull stations that require the operator to break a small piece of glass with a mallet are no longer recommended **(Figure 2.24)**. Originally, these devices were designed to discourage false alarms and were somewhat effective for that purpose. However, the broken glass presents an injury hazard to the operator at a time when an untrained operator is least capable of clear thinking. Although no longer sold, this type of pull station is still used in many older structures. A pull station may be protected by a wire or plastic "basket" in areas where it would be subject to damage or accidental activation **(Figure 2.25)**. This protective device may be found in gymnasiums, materials handling areas, or in locations where accidental activation presents a problem.

Figure 2.23 Pull boxes that are located near exits allow occupants to send an alarm as they are exiting the occupancy.

Figure 2.22 Most manual pull stations are red and white and have simple instructions for operation.

Figure 2.24 Older pull boxes that required the operator to break a glass are no longer recommended due to their potential for injury.

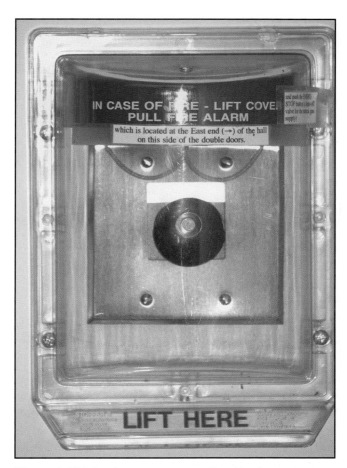

Figure 2.25 It is wise to protect a pull station in areas where it can be damaged or accidentally activated.

Coded Versus Noncoded Pull Stations

Depending on the type of system they serve, pull stations may be either coded or noncoded. Noncoded pull stations contain an on/off contact switch that is activated when the station is pulled. The pull station sends a continuous signal that sounds until the pull station and system are reset. Coded pull stations send an intermittent signal that indicates their location. Coded pull stations operate in a manner similar to that described earlier in the section on selective coded systems.

Single-Action Pull Station

These pull stations operate after a single motion is made by the user. When the station lever is pulled, a lever or other movable part is moved into the alarm position and a corresponding signal is sent to the system control unit **(Figure 2.26)**.

Double-Action Pull Station

As the name implies, these pull stations require the operator to perform two steps in order to initiate an alarm. First, the operator must lift a cover or open

Figure 2.26 A single-action pull station.

Figure 2.27 A double-action pull station requires that the operator lift a door and then activate the alarm.

Figure 2.28 Detection devices are available to recognize any of the products of combustion.

a door to access the alarm control **(Figure 2.27)**. Once this is done, the alarm lever, switch, or button must be operated to send the signal to the fire alarm control unit. These devices come in different colors and shapes, depending on the design of the manufacturer. Double-action pull stations may be confusing to certain occupant/operators due to the need to perform two separate steps before an alarm is initiated. Many people mistakenly believe that performing the first step activates the alarm.

Automatic Alarm-Initiating Devices

Automatic alarm-initiating devices, sometimes called detectors, are devices that continuously monitor the atmosphere of a given area. When certain changes in the atmosphere are detected, such as a rapid rise in heat or presence of smoke, a signal is sent to the fire alarm control unit **(Figure 2.28)**. These devices are typically very accurate at sensing the presence of the combustion products that they are designed to detect. It must be remembered, however, that many times these products can be found in a given area even when there is no emergency condition. For example, flame detectors may trip if a welder strikes an arc in a monitored area or smoke detectors may activate due to excessive moisture. These possibilities force fire protection system designers to take into account the normal activities that take place in

any given protected area. They then must design a detection system that minimizes the chances of an accidental activation.

There are four basic types of automatic alarm-initiating devices. They include those that detect heat, smoke, fire gases, and flame. The following sections describe the various types of devices in use.

Fixed-Temperature Heat Detector

Systems using heat detection devices are among the oldest types of fire detection systems. They are relatively inexpensive compared to other types of systems and they are the least prone to false activations. They are limited, however, by the fact that they are typically the slowest type of system to activate under fire conditions.

To be effective, heat detectors must be on ceilings, atriums, and architecturally designed areas where heat is expected to accumulate **(Figure 2.29)**. Detectors must also be selected at a tempera-

Figure 2.29 Heat detectors must be placed where heat is expected to accumulate.

ture rating that will give at least a small margin of safety above the normal ceiling temperatures that can be expected in that area. According to NFPA 72, heat-sensing fire detectors must be color-coded and marked with their listed operating temperatures. **Table 2.2** lists colors and temperatures for these detectors.

Because heat is a product of combustion, it is detectable by certain devices using three primary principles of physics:

- Heat causes expansion of various materials.
- Heat causes melting of certain materials.
- Heated materials have thermoelectric properties that are detectable.

All heat detection devices operate on one or more of these principles. The various types of fixed-temperature devices used in fire detection systems are covered in the following sections.

Fusible Links/Frangible Bulbs

Although fusible links and frangible bulbs are more commonly associated with automatic sprinklers, they are also used in fire alarm systems. The operating principles of links and bulbs that are used in fire detection and signaling systems are identical to the links and bulbs used with automatic sprinklers; it is their application that differs.

NOTE: Operations with automatic sprinklers are described in Chapter 6 of this manual.

Fusible links are used to hold a spring device in the detector in the open position (**Figure 2.30, p. 68**). When the melting point of the fusible link is reached, it melts and drops out. This causes the spring to release and touch an electrical contact that completes the circuit and sends an alarm signal. In order to restore the detector, the fusible link must be replaced.

A frangible bulb is designed so that it will fail at a specific temperature. The bulb is inserted into the detection device to hold two electrical contacts apart, much like that described for the fusible link. As the temperature increases, the liquid in the bulb expands, compresses the air bubble in the glass, and the bulb fractures and falls away. The contacts

Temperature Classification	Temperature Rating Range		Maximum Ceiling Temperature		Color Code
	°C	°F	°C	°F	
Low*	39 – 57	100 – 134	28	80	Uncolored
Ordinary	58 – 79	135 – 174	47	115	Uncolored
Intermediate	80 – 121	175 – 249	69	155	White
High	122 – 162	250 – 324	111	230	Blue
Extra high	163 – 204	325 – 399	152	305	Red
Very extra high	205 – 259	400 – 499	194	380	Green
Ultra high	260 – 302	500 – 575	249	480	Orange

**Table 2.2
Temperature Classification for
Heat-Sensing Fire Detectors**

* Intended only for installation in controlled ambient areas. Units shall be marked to indicate maximum ambient installation temperature.

Reprinted with permission from NFPA 72, *National Fire Alarm Code*. Copyright© 2002 National Fire Protection Association, Quincy MA 02269. This reprinted material is not the complete and official position of the National Fire Protection Association on the referenced subject, which is represented only by the standard in its entirety.

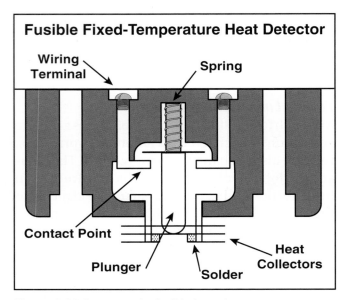

Fusible Fixed-Temperature Heat Detector

Wiring Terminal

Spring

Contact Point

Plunger

Solder

Heat Collectors

Figure 2.30 Cutaway of a fusible heat detector.

close to complete the circuit and send the alarm. In order to restore the detector, either the frangible bulb or the entire detector must be replaced.

Continuous Line Detector

Most of the detectors described in this chapter are the "spot" detector type; that is, they detect conditions only at the spot where they are located. However, one type of heat detection device — the continuous line device — can be used to detect conditions over a wide area.

One type of continuous line detection device consists of a conductive metal inner core cable that is sheathed in a stainless steel tubing (**Figure 2.31**). The inner core and the sheath are separated by an electrically insulating semiconductor material.

This material keeps the core and the sheath from touching but allows a small amount of current to flow between them. This insulation is designed to lose some of its electrical resistance capabilities at a predetermined temperature anywhere along the line. These cables can be strung out over extremely large areas. When the heat at any given point reaches the resistance reduction point of the insulation, the amount of current transferred between the two components increases. This results in an alarm signal being sent to the system control unit. This detection device restores itself when the level of heat is reduced.

Another type of continuous line detection device uses two wires that are each insulated and bundled within an outer covering. When the melting temperature of each wire's insulation is reached, the insulation melts and allows the two wires to touch. This completes the circuit and sends an alarm signal to the fire alarm control unit (**Figure 2.32**). To restore this type of line detector, the fused portion of the wires must be cut out and replaced with new wire.

Bimetallic Detector

A bimetallic detector uses two types of metal that have different heat-expansion ratios. Each metal is formed into thin strips and the different metals are then bonded together. The fact that one metal expands faster than the other will cause the combined strip to arch when subjected to heat. The amount that it arches depends on the characteristics of the metals, the amount of heat it is exposed to, and the degree of arch present when in the

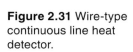

Figure 2.31 Wire-type continuous line heat detector.

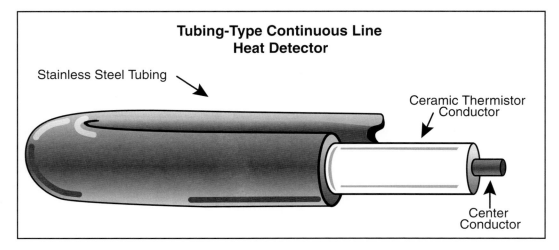

Tubing-Type Continuous Line Heat Detector

Stainless Steel Tubing

Ceramic Thermistor Conductor

Center Conductor

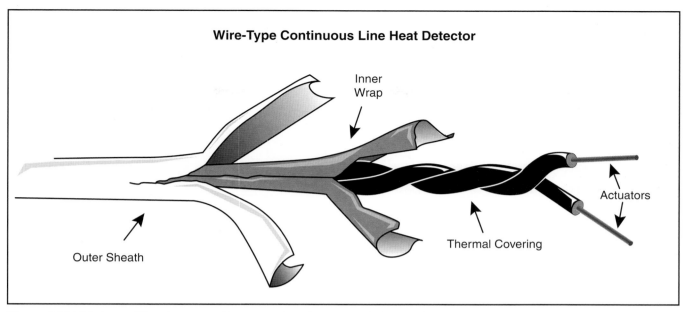

Wire-Type Continuous Line Heat Detector

Inner Wrap

Actuators

Thermal Covering

Outer Sheath

Figure 2.32 This type of line detector utilizes two wires that are each insulated and bundled within an outer covering.

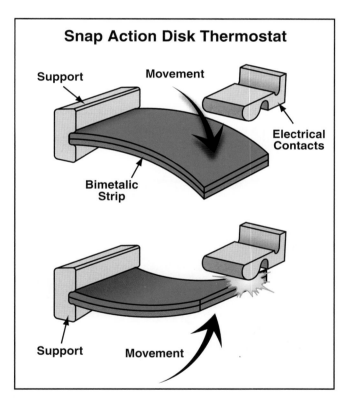

Snap Action Disk Thermostat

Support

Movement

Electrical Contacts

Bimetalic Strip

Support

Movement

Figure 2.33 A bimetallic heat detector in the normal and activated positions.

normal position. All of these factors are calculated into the design of the detector.

The bimetallic strip may be positioned with either one or both ends secured in the device. When positioned with both ends secured, a slight bow is placed in the strip. When heated, the expansion causes the bow to snap in the opposite direction (**Figure 2.33**). Depending on the design of the device, this action either opens or closes a set of electrical contacts that in turn send an alarm signal to the fire alarm control unit. Most bimetallic detectors are the automatic resetting type. They do need to be checked, however, to ensure that they have not been damaged.

Rate-of-Rise Heat Detector

A rate-of-rise heat detector operates on the principle that fires rapidly increase the temperature in a given area. The rate-of rise detector is designed to detect these quick increases in temperature. These detectors respond at substantially lower temperatures than fixed-temperature detectors. Typically, rate-of-rise heat detectors are designed to send a signal when the rise in temperature exceeds 12°F to 15°F (7°C to 8°C) per minute. This is because temperature changes of this magnitude are not expected under normal, nonfire circumstances.

Most rate-of-rise heat detectors are reliable and are not subject to false activations. However, they can occasionally be activated under nonfire conditions. An example of this would be when a rate-of rise detector is placed near a garage door in an air-conditioned building. If the garage door is opened on a hot day, the influx of heated air will rapidly increase the temperature around the detector, causing it to activate. These situations can be avoided by proper placement. There are several different types of rate-of-rise heat detectors in use; all are automatically reset. The following types are discussed in more detail in the following sections:

- Pneumatic rate-of-rise spot detector
- Pneumatic rate-of-rise line detector
- Rate compensated detector
- Thermoelectric detector

Pneumatic Rate-of Rise Spot Detector

The pneumatic spot detector is the most common type of rate-of-rise detector in use (**Figures 2.34 a and b**). It contains a small chamber that is filled with air and has a flexible metal diaphragm in the bottom. As the air inside the chamber expands and the temperature rises, the diaphragm is forced out to a predetermined level. Depending on the design of the particular detector, this movement causes a set of electrical contacts to either open or close, thus sending an alarm signal to the fire alarm control unit.

The air chambers in these detectors have to be vented to prevent activation caused by normal changes in ambient temperature or changes in barometric pressure. The vent is designed to be of a size that allows air to enter or exit the chamber at a predetermined pressure below that which will activate the detector.

Pneumatic Rate-of-Rise Line Detector

The spot detector monitors its exact area of location; however, a line detector can monitor large areas. Line detectors consist of a system of tubing arranged over a wide area of coverage (**Figure 2.35**). The space inside the tubing acts as the chamber that was described previously in the section on spot detectors. These detectors also contain a diaphragm and are vented. When an area being served by the tubing experiences a temperature increase, the detector functions in the same manner as that described for the spot detector.

The tubing in these systems must be limited to about 1,000 feet (300 m) in length. The tubing should be arranged in rows that are not more than 30 feet (10 m) apart and 15 feet (5 m) from walls.

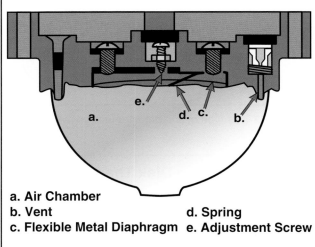

a. Air Chamber
b. Vent
c. Flexible Metal Diaphragm
d. Spring
e. Adjustment Screw

Figures 2.34 a and b The pneumatic spot detector is the most common type of rate-of-rise detector.

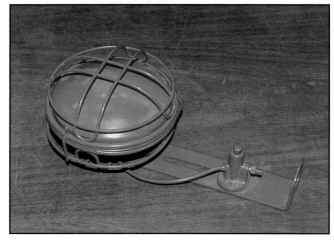

Figure 2.35 A pneumatic rate-of-rise line detector.

Rate Compensated Detector

This detector is designed for use in areas that are subject to regular temperature changes, but at rates that are slower than fire conditions. Rate compensated detectors contain an outer bimetallic

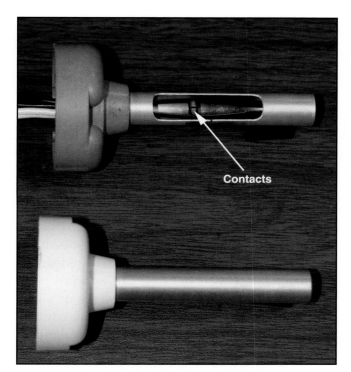

Figure 2.36 In a rate-compensated detector, rising temperatures cause the electrical contacts inside the detector to touch, thus sending an alarm.

sleeve with a moderate expansion rate. This outer sleeve contains two bowed struts that have a slower expansion rate than the sleeve **(Figure 2.36)**. The bowed struts have electrical contacts on them. In the normal position these contacts do not come together. When the detector is heated rapidly, the outer sleeve expands in length. This reduces the tension on the inner strips and allows the contacts to come together, thus sending an alarm signal to the fire alarm control unit.

If the rate of temperature rise is fairly slow, such as 5 to 6 degrees Fahrenheit (or 2 to 3 degrees Celsius) per minute, the sleeve expands at a slow rate that will maintain tension on the inner strips. This prevents unnecessary system activations.

Thermoelectric Detector

This rate-of-rise detector operates on the principle that two wires made of dissimilar metals, when twisted together and heated at one end, will cause an electrical current to be generated at the other end. The rate at which the wires are heated determines the amount of current that is generated. These detectors are designed to bleed off or dissipate small amounts of current, which reduces the

chance of a small temperature change activating an alarm. Greater changes in temperature result in larger amounts of current flowing and in activation of the alarm system.

Smoke Detector

Most people are well aware of the dangers of fire but less aware of the dangers of smoke inhalation. About 65 percent of fire deaths are attributed to smoke inhalation, not burns as many people think. Smoke and deadly gases spread farther and faster than the heat from flames. When people are asleep, deadly fumes can quickly send them into a deeper level of unconsciousness. Because of these dangers, an early warning can mean the difference between a safe escape and no escape at all.

A smoke detector senses the presence of a fire much more quickly than does a heat detection device. The smoke detector is the preferred detector in many types of occupancies and is used extensively in residential, institutional, and health care settings. Many factors affect the performance of smoke detectors of all types, such as the type and amount of combustibles, the rate of fire growth, the proximity of the detector to the fire, and ventilation within the area involved.

Smoke detectors are listed by their performance in fire tests. Regardless of their principle of operation, all smoke detectors are required to respond to the same fire tests. There are two basic types of smoke detectors in use: photoelectric and ionization. The allowable sensitivity ranges for both types of smoke detectors are established by Underwriters Laboratories Inc.

Photoelectric Smoke Detector

A photoelectric detector, sometimes referred to as a visible products-of-combustion smoke detector, uses a photoelectrical cell coupled with a specific light source. The photoelectric cell functions in either of two ways to detect smoke: projected beam application (obscuration) and refractory application (scattered).

The projected beam application type of photoelectric detector uses a beam of light focused across the area being monitored onto a photoelectric receiving device, such as a photodiode. The cell constantly converts the beam into current, which keeps a switch open. When smoke interferes with the light beam, the amount of current produced is lessened. The change in current output is sensed by the detector circuitry, and when a threshold is crossed an alarm is initiated. Obscuration detectors are usually of the projected beam type where the light source expands an area to be protected (**Figure 2.37**).

Projected beam smoke detectors are particularly useful in areas where a large area of coverage is desired, such as in churches, atriums, or warehouses. Rather than wait for smoke particles to collect at the top of an open area and sound an alarm, the beam-type detector is strategically positioned to sound an alarm more quickly. Beam smoke detectors consist of a transmitter that projects an infrared beam across the protected area and a receiver containing a photosensitive cell. The photosensitive cell then monitors the signal strength of the light beam. If the light beam is interrupted by smoke particles or any other visible medium, an alarm will sound. With beam smoke detectors, it is important that they be mounted on a stable stationary surface. Any movement due to temperature variations, structural movement, and vibrations can cause the light beams to become misaligned.

The refractory photocell uses a beam of light from a light-emitting diode (LED) that passes through a small chamber at a point away from the light source. Normally, the light does not strike the photocell or photodiode. When smoke particles enter the light path, light strikes the particles and is reflected in random directions onto the photosensitive device, causing the detector to respond with an alarm signal (**Figure 2.38**).

A photoelectric smoke detector can work satisfactorily on all types of fires, but generally it responds more quickly to smoldering fires than ionization detectors do. Photoelectric smoke detectors are best suited for living rooms, bedrooms, and kitchens. This is because these rooms often contain such large pieces of furniture as sofas, chairs and mattresses that can burn slowly and create more smoke than flames. This detector is automatically reset.

Ionization Smoke Detector

An ionization type smoke detector contains a sensing chamber consisting of two electrically charged plates and a radioactive source for ionizing the air between the plates. A small amount of radioactive material — Americium 241 – that is adjacent to the opening of the chamber ionizes the air particles as they enter. Inside the chamber are two electrical plates: one positively charged and one negatively

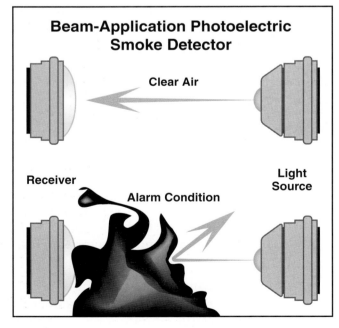

Figure 2.37 Principle of a beam-application smoke detector.

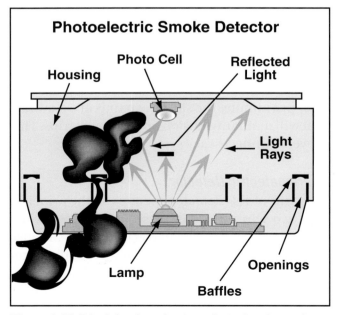

Figure 2.38 Principle of a refractory photoelectric smoke detector.

charged. The ionized particles free electrons from the negative plate, and the electrons travel to the positive plate. Thus, a small ionization current measurable by electronic circuitry flows between the two plates. Products of combustion, which are much larger than the ionized air molecules, enter the chamber and collide with the ionized air molecules. As the two interact, they combine and the total number of ionized particles is reduced. This results in a decrease in the chamber current between the plates. When a certain threshold is crossed an alarm is initiated (**Figure 2.39**).

To compensate for the possible effects of humidity and pressure changes, a dual ionization chamber detector was developed and is found in most jurisdictions. A dual chamber detector uses two ionization chambers. One chamber senses particulate matter, humidity, and atmospheric pressure. The other chamber is a reference chamber that is partially closed to outside air and is affected only by humidity and atmospheric pressure. Both chambers are monitored electronically and their outputs are compared. If the humidity or atmospheric pressure changes, both chambers are affected equally and cancel each other. When particles of combustion enter the sensing chamber,

its current decreases while the reference chamber remains unchanged. The imbalance in current is detected electronically and an alarm is initiated.

An ionization detector works satisfactorily on all types of fires; however, it generally responds quicker to flaming fires than do photoelectric smoke detectors. Ionization smoke detectors are best suited for rooms that contain highly combustible materials, such as the following:

- Cooking fat/grease
- Flammable liquids
- Newspapers
- Paint
- Cleaning solutions

This detector is an automatic resetting type.

Air-Sampling Smoke Detector

An air-sampling smoke detector is a type of ionization smoke detector that is designed to continuously monitor a small amount of air from the protected area for the presence of smoke particles. There are two basic types of air-sampling smoke detectors. The most common type is the cloud chamber type (**Figure 2.40**). This detector uses a small air pump to draw sample air into a high-humidity chamber within the detector. The detector then imparts the high humidity to the sample and lowers the pressure in the test chamber. Moisture condenses on any smoke particles in the test chamber, which

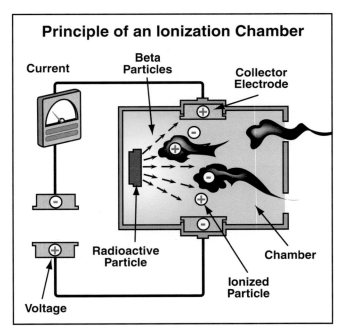

Figure 2.39 Products of combustion interfere with the ionized particles and result in a decrease in the chamber current between the plates, eventually causing an alarm to sound.

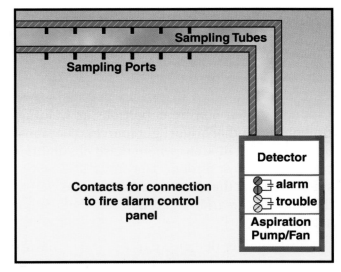

Figure 2.40 The cloud chamber air sampling smoke detector is the most common type of air-sampling detector.

creates a cloud in the chamber. The detector triggers an alarm signal when the density of this cloud exceeds a predetermined level.

The second type of air-sampling smoke detector is comprised of a system of pipes spread over the ceiling of the protected area (**Figure 2.41**). A fan in the detector/controller unit draws air through the pipes. The air is then sampled using a photoelectric sensor.

Limitations of Smoke Detectors

Smoke detectors offer the earliest possible warning of fire. They have saved thousands of lives already and will continue to do so. Nonetheless, smoke detectors have certain limitations:

- They may not provide early warning of a fire developing on another level of a building.
- They may not detect fire developing on the other side of a closed door.
- They may not be effective when fire is caused by explosions resulting from careless housekeeping, such as improper storage of hazardous materials or flammable liquids.

Flame Detector

A flame detector is sometimes called a light detector. There are three basic types:

- Those that detect light in the ultraviolet wave spectrum (UV detectors) (**Figure 2.42 a**).
- Those that detect light in the infrared wave spectrum (IR detectors) (**Figure 2.42 b**).
- Those that detect both UV and IR waves.

While these types of detectors are among the fastest to respond to fires, they are also easily tripped by such nonfire conditions as welding, sunlight, and other bright light sources. They must only be placed in areas where these possibilities can be avoided or limited. They must also be positioned so that they have an unobstructed view of the protected area. If they are blocked, they cannot activate.

An ultraviolet detector can give false alarms when it is in contact with sunlight and arc welding, so it must be placed where these and other sources of ultraviolet light can be eliminated. An infrared

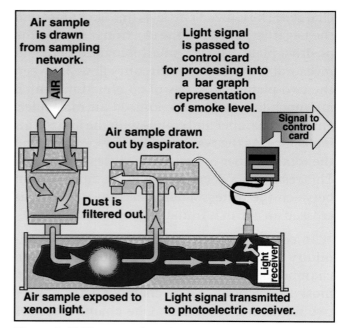

Figure 2.41 The operating principles of a tube-type air sampling smoke detector.

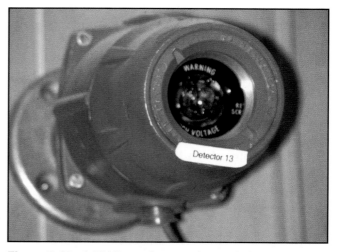

Figure 2.42a A UV flame detector detects light in the ultraviolet wave spectrum.

Figure 2.42b A typical infrared flame detector. *Courtesy of Detector Electronics Corp.*

detector is effective in monitoring large areas such as an aircraft hangar or computer room **(Figure 2.43)**. To prevent accidental activation from infrared light sources other than fires, an infrared detector requires the flickering action of a flame before it activates to send an alarm. This detector is typically designed to respond to 1 square foot (.3m²) of fire from a distance of 50 feet (15 m).

Fire Gas Detector

When fire breaks out in any confined area, it drastically changes the chemical gas content of the atmosphere in that area. Some of the gases released by a fire may include:

- Water vapor (H_2O)
- Carbon dioxide (CO_2)
- Carbon monoxide (CO)
- Hydrogen chloride (HCl)
- Hydrogen cyanide (HCN)
- Hydrogen fluoride (HF)
- Hydrogen sulfide (H_2S)

Only water, carbon dioxide, and carbon monoxide are released from all carbonaceous materials that burn. Whether other gases are released depends on the specific chemical makeup of the fuel, so it is only practical to monitor levels of carbon dioxide and carbon monoxide for fire detection purposes. A fire gas detector operates somewhat faster than a heat detector but not as fast as a smoke detector. It uses either semiconductors or catalytic elements to sense a particular gas or gases and

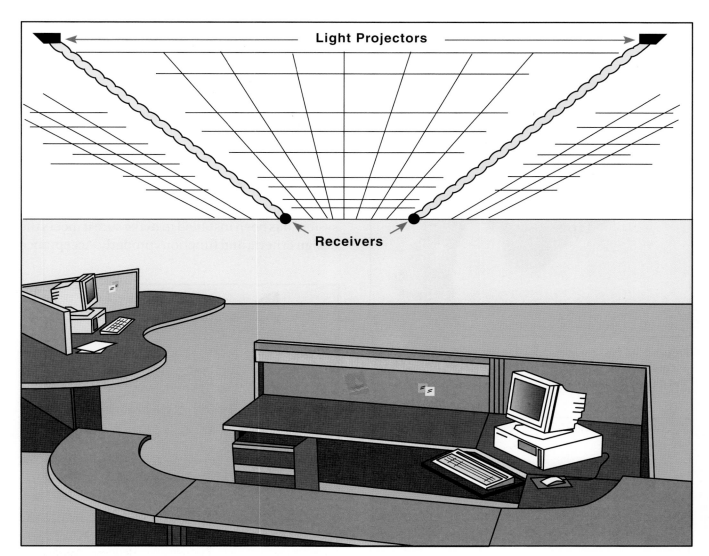

Figure 2.43 Infrared flame detectors can be used to protect large areas.

trigger the alarm. Fire gas detectors are not used as frequently as other types of detectors but can be found in such places as refineries, chemical plants, and areas of electronic assembly **(Figure 2.44)**.

Combination Detector

Depending on the design of the system, various combinations of the previously described detection devices may be used in a single device. These combinations include fixed rate/rate-of-rise detectors, heat/smoke detectors, and smoke/fire gas detectors **(Figure 2.45)**. These combinations give the detector the benefit of both services and increase their responsiveness to fire conditions.

Figure 2.44 Fire gas detectors use either semiconductors or catalytic elements to sense a particular gas or gases and trigger the alarm. *Courtesy of RKI Instruments, Inc.*

Figure 2.45 A combination heat/smoke detector.

Inspecting and Testing Fire Detection and Signaling Systems

In order to ensure operational readiness and proper performance, fire detection and signaling systems must be tested when they are installed and again on a continuing basis. Tests conducted when systems are installed are commonly called acceptance tests. Periodic testing is often referred to as service testing.

Fire department and fire brigade personnel who routinely conduct inspections need to have a working knowledge of these systems **(Figure 2.46)**. It is important to keep in mind, however, that these personnel are generally limited to visual inspections and to supervision of system tests. They will not have to operate or maintain these systems. In most cases, in-plant personnel or alarm system contractors actually perform system tests and maintenance. Personnel performing inspection, testing, and maintenance should be qualified and experienced in the types of devices and systems they deal with.

NOTE: Refer to NFPA 72 for more information.

Acceptance Testing

Acceptance testing is performed soon after the system has been installed to make sure it meets the design criteria and functions properly. Acceptance

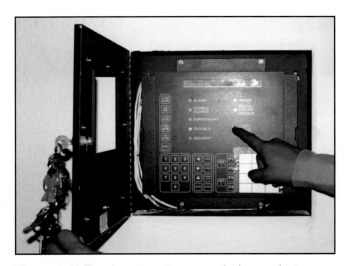

Figure 2.46 Fire department personnel who conduct inspections must have a working knowledge of detection and alarm systems.

tests may be required by the occupant's insurance carrier and/or local codes and ordinances. These tests should be witnessed by representatives of the building owner, the fire department, and the system installer/manufacturer. The fire department representative may be a fire inspector, or a staff fire protection engineer, or in some cases the fire marshal. Some jurisdictions require a test certificate from the system manufacturer/installer indicating that the system has been thoroughly tested before the fire department inspection. This prevents the fire inspector from spending time checking a system that may not have been properly installed.

All functions of the fire detection and signaling system should be operated during the acceptance tests:

- Both the alarm and trouble modes of system operation should be tested. Actual wiring and circuitry should be checked against the system drawing to ensure that all are connected properly.

- The system control panel should be thoroughly inspected to ensure that it is in proper working order (**Figure 2.47**). All interactive controls at the panel should be operated to ensure that they control the system as designed.

- All alarm-initiating and indicating devices and circuits must be checked for proper operation. Pull stations, detectors, bells, strobe lights etc., should be tested to make sure they are operational. Each initiating device must be tested to ensure that it sends an appropriate signal and causes the system to go into the alarm mode.

- The system should be operated on both the primary and secondary power supplies to make sure both will supply the system adequately.

Restorable heat detectors should be checked by following approved testing procedures described by the manufacturer (**Figure 2.48**). **Never allow an open flame to be used for testing detectors.** There is an obvious fire hazard associated with this practice along with the potential for damage to the detector itself. Heat detectors are designed to detect changes in temperature, not fire. Remember that some combination detectors have both restorable and nonrestorable elements. Exercise

Figure 2.47 It is very important to verify that the system control panel is in working order.

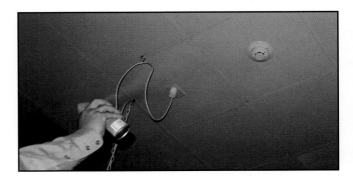

Figure 2.48 It is important to use approved testing procedures for restorable heat detectors.

caution to avoid tripping the nonrestorable element. Nonrestorable pneumatic detectors should be tested mechanically. Those detectors equipped with replaceable fusible links should have the links removed to see whether the contacts touch and send an alarm signal. The links can then be replaced.

The manufacturers of smoke, flame, and fire gas detectors usually have specific instructions for testing their detectors. These instructions must be followed on both the acceptance and the service tests. They may include the use of smoke-generating devices, aerosol sprays, or magnets. The use of nonapproved testing devices may result in the manufacturer's warranty on the detector being voided.

It is also important to check the response of outside entities that are responsible for monitoring the system. This is important on central station, auxiliary, remote station, and proprietary systems. The alarm receiving capability and response of those involved must be verified, and a 9-1-1 dispatch facility must receive the request for emergency service.

The results of all tests must be documented to the satisfaction of both the insurance carrier and the fire department. Only after all parts of the system have successfully passed the tests should a system certification certificate be issued. This is typically a preliminary step toward the issuance of a *certificate of occupancy*. NFPA 72 contains complete information on system acceptance tests.

System Service Testing and Periodic Inspections

Fire detection and signaling equipment should receive a general inspection on a routine basis. The inspections should be conducted by both the fire authority having jurisdiction and by the building owner/occupant. Because fire departments may have many occupancies in their jurisdiction and it is time-consuming to test these systems, it is not possible for the fire department to participate at every test. Most of the time, occupants will have to test the systems on their own and document the results. At intervals specified by the fire department, fire department personnel will also witness the tests. The procedures covered in this section should be applied during occupant and/or fire department inspections and testing.

NOTE: Testing and inspection intervals may vary in accordance with local codes and ordinances.

Inspect all wiring for proper support, wear, damage, or any other defects that may render the insulation ineffective. Where circuits are enclosed

in conduit, inspect the conduit for solid connections and proper support. When batteries are used as an emergency power source, they should be checked for clean contacts and proper charge. Many batteries have floating-ball indicators that show whether they are properly charged. In accordance with NFPA 72, lead-acid and primary (dry-cell) type batteries should be inspected monthly and nickel-cadmium and sealed lead-acid types should be inspected semi-annually. Batteries that fail the inspection and testing procedure should be replaced immediately.

All equipment, especially initiating and indicating devices, must be kept free of dust, dirt, paint, and other foreign materials. When either dust or dirt is found, it is recommended that the devices be cleaned with a vacuum cleaner rather than by wiping (**Figure 2.49**). Wiping tends to

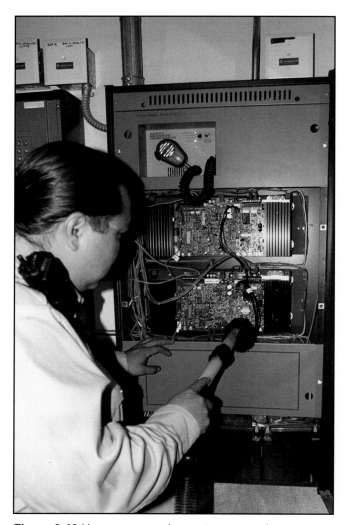

Figure 2.49 Use a vacuum cleaner to remove dust.

spread debris around, causing it to settle on electrical contacts. This may inhibit the future operation of the system.

System control units, recording instruments, and other devices should not have objects stored on, in, or around them. Many system control panels that are designed with locking doors have storage areas for extra relays, light bulbs, and test equipment. If this space is not designed into the unit, these devices should be stored somewhere else. Otherwise, they may foul moving parts or cause electrical shorts that can result in system failure.

Inspecting and Service Testing Initiating Devices

Any detection and signaling system will be ineffective unless the alarm-initiating devices are in proper working order and send the appropriate signal to the system control panel. The following sections highlight the procedures for checking these devices.

Manual alarm-initiating devices. Numerous items need to be checked when testing and inspecting a manual alarm-initiating device. Access to the device must be unobstructed and each unit should be easy to operate. The housing should be tightly closed to prevent dust and moisture from entering the unit and disrupting service **(Figure 2.50)**. Any chipped, cracked, or otherwise impaired glass should be removed and replaced. If the device is equipped with a cover or door, it should be checked to make sure it opens easily and all the components behind the door are in place and ready for service. Inspectors may wish to witness the activation of selected devices to ensure that the device and the system are operational.

Automatic alarm-initiating devices. Without functional detection devices, the most elaborate wiring and signaling systems are useless. The reliability of the entire system is, in fact, based largely on the reliability of the detection devices. Automatic alarm-initiating devices should be checked after installation, after a fire, and periodically based on guidelines established by state or local authorities or the manufacturer. Often these guidelines are found in the fire code that has been adopted by local agencies. All detector testing should be in accordance with local guidelines, manufacturer's specifications, and NFPA 72.

Figure 2.50 Secure the alarm system housing to keep out dirt and dust.

This section highlights the procedures and issues related to detector inspections and testing.

Detectors must not be damaged or painted. Regardless of the type of detector in use, the following detectors should be replaced or sent to a recognized testing laboratory for testing:

- Detectors on systems that are being restored to service after a period of disuse
- Detectors that are obviously corroded
- Detectors that have been painted over, even if attempts were made to clean them
- Detectors that have been mechanically damaged or abused
- Detectors on circuits that were subjected to current surges, overvoltages, or lightning strikes
- Detectors subjected to foreign substances that might affect their operation
- Detectors subjected to either direct flame, excessive heat, or smoke damage

A permanent record of all detector tests must be maintained for at least five years. The minimum information that should be included in the record is the date, the detector type, the location, the type of test, and the results of the test.

A nonrestorable fixed-temperature detector cannot be tested periodically. Testing would destroy the detector and require the system to be rendered inoperable until a replacement detector could be located and installed. For this reason, tests are not required until 15 years after the detector has been installed. At this time, 2 percent of the detectors must be removed and laboratory tested. If a failure occurs in one of the detectors, additional detectors must be removed and laboratory tested. These tests are designed to determine if there is a problem with failure of the product in general or a localized failure involving just one or two detectors. Line detectors must have resistance testing performed semiannually.

A restorable heat detection device should be checked as described previously in the acceptance testing section. One detector on each signal circuit should be tested semiannually. A different detector should be selected each time and so noted on the inspection report. Subsequent inspections should include a copy of the previous report to ensure that the same detector is not tested each time.

A fusible-link detector with replaceable links should also be checked semiannually. This is done by removing the link and observing whether or not the contacts close **(Figure 2.51)**. After the test, the fusible link must be reinstalled. It is recommended that the links be replaced at five-year intervals.

A pneumatic detector should be tested semiannually with a heating device or a pressure pump. If a pressure pump is used, the manufacturer's instructions must be followed.

A smoke detector should be tested semiannually in accordance with manufacturer's recommendations. The instruments required for performance and sensitivity testing are usually provided by the manufacturer. Sensitivity testing should be performed after the detector's first year of service and every two years after that. **Blowing cigarette smoke into the detector is *not* an acceptable method of testing smoke detectors.**

Figure 2.51 It is important to remove the link in a fusible-link detector and observe whether or not the contacts close.

Flame and gas detection devices are very complicated devices and should only be tested by highly trained individuals. Testing is typically performed by professional alarm service technicians on a contract basis.

Inspecting System Control Units

Inspectors should check the system control panel to ensure that all parts are operating properly. All switches should perform their intended functions and all indicators should light or sound when tested **(Figure 2.52)**. When individual detectors are triggered, the system control unit should indicate the proper location and the warning lamps should light. It is important to remember that the indicated location could very well be out of date due to renovations. Auxiliary devices can also be checked at this time. The auxiliary devices include the following: local evacuation alarms, HVAC functions such as air-handling system shutdown controls, fire dampers, and the like. All devices must be restored to proper operation after testing.

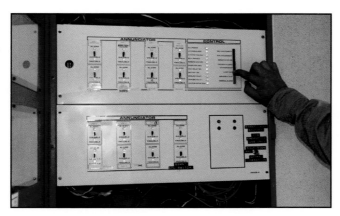

Figure 2.52 All parts of a control panel should light or sound when tested.

In connection with these tests, the receiving signals should also be checked. The proper signal and/or number of signals should be received and recorded. Signal impulses should be definite, clear, and evenly spaced to identify each coded signal. There should be no sticking, binding, or other irregularities. At least one complete round of printed signals should be clearly visible and unobstructed by the receiver at the end of the test. The time stamp should clearly indicate the time of the signal and should not interfere in any way with the recording device.

System Testing Timetables

The following sections give a brief synopsis of the inspection and testing requirements for various types of systems. If any of these systems use backup electrical generators for emergency power, those generators should be run under load monthly for at least 30 minutes.

Local systems. Local alarm systems should be tested in accordance with guidelines established by the authority having jurisdiction and/or the occupant. There are no NFPA standard requirements specifying the test period, but a fire department should develop its own requirements for periodic testing of local systems. It is important for fire department personnel to be both aware of and familiar with all alarm systems within their jurisdiction.

Central station systems. Central station signaling equipment should be tested on a monthly basis. Waterflow indicators, automatic fire detection systems, and supervisory equipment should be checked bimonthly. Manual fire alarm devices, water tank level devices, and other automatic sprinkler system supervisory devices should be checked semiannually. When these tests are to be conducted, it is important that both plant supervisory personnel and central station personnel be notified before the test. This will prevent them from dispatching fire units or evacuating occupants.

Auxiliary systems. Auxiliary fire alarm systems should be visually inspected and actively tested monthly by the occupant to see that all parts are in working order and that the operation of the system results in a signal sent to the fire telecommunications (dispatch) center. Noncoded manual fire alarm boxes should be tested semiannually.

Remote station and proprietary systems. Most of the testing requirements for these types of systems are established by the authority having jurisdiction. The fire detection components of these systems should be tested monthly. Waterflow indicators should be tested semiannually; however, the frequency of testing may depend upon the type of indicator.

NOTE: In all cases of testing and inspection frequencies, check with the most current edition of NFPA 72, *National Fire Alarm Code®*.

Emergency voice/alarm systems (notification appliances). A functional test of the various components in these systems should be conducted on a quarterly basis by the occupant. These tests can include selected parts of the system that are reflective of what may actually be used during an incident. However, all components must be checked at least annually.

Record Keeping

The following section on record keeping is fundamental information used in most facets of the business world. The paperwork and quality of records can be very important in resolving disputes with the public or in a court of law, so all personnel involved should be aware of the importance of accurate and complete records. When conducting fire inspections, the readiness of fire suppression and detection systems must be documented and stored systematically. In the performance of an occupation centered on public safety, good record keeping is good business.

One of the most crucial functions carried out by any fire agency is that of maintaining accurate files and records of all occupancies it is responsible for inspecting. Most often these records are maintained by the fire alarm contractor and/or the building owner, and provide a historical perspective of fire prevention activities within that jurisdiction. They also provide the basis for all future code compliance and enforcement activities. Accurate record keeping is important in any business, and never more so than when records have been subpoenaed into a court of law. In litigation, a single carelessly prepared document can potentially cost millions of dollars in judgments against a fire agency.

All documents and records pertaining to code enforcement activities should be kept within the agency's document management system. These include the following:

- Inspection reports, forms, and letters
- Violation notices and citations
- Court summonses
- Plans review comments, approvals, and drawings (**Figure 2.53**)
- Fire reports
- Fire investigations
- Compliant permits and certificates issued

It is most desirable to maintain files for all properties (excluding one- and two-family residences) within the jurisdiction. Files should be maintained on occupancies that have the following characteristics:

- Have been issued a permit, certificate, or license of some type
- Contain automatic fire suppression or detection systems
- Conduct hazardous operations or house hazardous materials on a routine basis
- Have had a fire incident

It is recommended that records be maintained on a building for the entire life of that building. In other words, if the building is still standing, there should be records kept on that building and made readily available in the offices of the local fire agency. It is very common for a building to house many different owners and occupancies. By maintaining a file on the structure throughout its lifetime, inspectors will be able to note the changes that have been made to the structure and evaluate how proposed changes may be affected by the previous uses of the structure. If a structure is demolished, the fire prevention bureau should maintain the files for a reasonable period before the records and files are destroyed. Local policy and legal recommendations will dictate how long this period will be.

All files and records maintained by the fire prevention bureau are considered public domain documents. Obviously, some information contained in the reports may be considered proprietary or confidential in nature. Officials who are responsible for overseeing the storage of inspection records should understand the freedom of information laws and requirements for the jurisdiction they serve. This will provide guidance for them on what documents may be released to the public and those that must be withheld.

Written Records

Even in agencies that have been diligent in keeping up with technological advances, there tends to be a large number of written inspection records and documentation that needs to be filed and maintained. This includes older inspection records that were generated before computerization as well as hard copy documents that are now used in computer filing systems. Some agencies that have computerized their inspection documentation process have also chosen to print hard copies of the

Figure 2.53 Plans review comments and drawings are an important part of record-keeping.

information and maintain them in a written file. This ensures that new records are stored with older records to keep the file complete and consistent. These copies provide an additional backup in the event that a computer system failure ever makes the files inaccessible for any period of time.

Each inspected property should have a file that contains copies of all building and inspection records for that property (**Figure 2.54**). Each time an inspector has contact with the occupant or the property owner, records of those actions should be added to the file. It is important that the file be kept as up-to-date as possible.

The methods for cataloging and storing the files vary from agency to agency. The way file storage is handled depends on the size of the jurisdiction and the number of occupancies it is responsible for inspecting. Smaller agencies may be able to store such documents in simple filing cabinets. Larger agencies may have entire rooms dedicated

Figure 2.54 Written inspection records and documentation need to be filed and maintained in each facility.

to file storage. For agencies that must maintain large numbers of files and documents, older records may be reduced to microfilm to save storage space. This format is particularly helpful with such large documents as building plans or architectural drawings.

Because the occupants in a building tend to change over time, it is generally not wise to catalog files alphabetically by occupant name. The most reliable method of cataloging inspection records is by the building's street address. This will allow the file to stay in the same location even as building occupants change.

Electronic Records

Many fire departments and other inspection agencies are taking advantage of advances in computer technology to maintain and store inspection records. Computerized inspection record systems can also assist in planning and scheduling future inspection needs. The level of sophistication of the computer system is a reflection of the fire department or inspection agency's needs. Most departments use a purchased software programs; few departments have the resources to develop their own inspection data management system.

There are two primary methods by which data may be logged into the computer system:

- Inspectors use laptop computers or handheld electronic data recording equipment while making the inspection and then download the information into the system (**Figure 2.55, p. 84**).

- Inspectors use written forms to record the information while performing the inspection and then enter the information manually into the computer system upon returning to the office.

The ability to electronically record information in the field and then download it into the computer system is the more efficient of the two methods. As further advances in technology make small, portable computer equipment more accessible and affordable, this method will become a more common record-keeping practice.

There are many aspects of computer system management that must be given careful consideration when determining how to store inspection

Figure 2.55 Inspectors may use laptops to transfer data into the main computer system.

Fire inspectors who work for agencies that use computers for inspection records and data management must receive the appropriate amount of training on the system. Even the best-designed computer data management system will be useless unless the information it is designed to manage is entered and stored correctly.

The occupancy of a building may change many times. In record keeping for all fire alarm systems, it is important to evaluate the type of occupancy and the type of alarm system to ensure that they are appropriately matched. It is usually considered acceptable to use more restrictive requirements when installing a fire alarm system; conversely, it is unacceptable for a system to be less restrictive than NFPA 72, *National Fire Alarm Code®*, or local codes and ordinances permit.

Summary

Whether a fire detection and alarm system is simple or complex, its quick operation is vital if fire emergencies are to be mitigated with as little loss or damage as possible. The earliest systems of ringing bells and crying out warnings have given way to complex electronic systems that alert occupants and emergency services at the same time. Fire fighting personnel must be familiar with the basic operating features of the detection and alarm systems in their response areas so that they can recognize obvious problems that may be occurring. They also need to know which types of detection systems are appropriate for certain hazards so that the systems that are installed can be as effective as they were intended. Because this level of inspection is confined to visual observations and supervision of systems testing, it is also important to involve other trained personnel to ensure system operability and safety.

records and data. Several questions that must be answered are as follows:

• How will the information be filed?

• How can the information be retrieved?

• What portion of the information will be stored in a read-only format so that records cannot be accidentally or purposely changed without authorization?

• Which personnel will be given access to retrieve information from the system?

• What information can be released to the public?

Introduction to Water Supply

1. Explain the extinguishing properties of water.

2. Describe the advantages and disadvantages of using water as an extinguishing agent.

3. Describe different types of pressure as they relate to water.

4. List the factors and conditions that contribute to friction loss in fire hose and pipes.

5. List the basic components of a municipal water supply system.

6. Describe how a water treatment facility can affect the availability of water for fire extinguishment.

7. Describe the fundamental components of a water distribution system.

8. Describe how fire hydrants are located and distributed according to the uniform fire codes.

9. Explain the three purposes for the use of a private water supply system.

10. Describe the water supply requirement for a standpipe and hose system.

11. Explain the water supply requirements for an automatic fire sprinkler system.

12. Describe the two methods used for the planning and design of automatic fire sprinkler systems.

FESHE Objectives

Fire and Emergency Services Higher Education (FESHE) Objectives:
Fire Science Curriculum: Fire Protection Systems

- Articulate knowledge of distribution and installation of water supply systems in suburban and rural areas.

Chapter 3
Introduction to Water Supply

Water remains the most abundant and available resource for combating fire. For these reasons, the advantages and disadvantages of water as an extinguishing agent and the difficulty of obtaining and moving water must be thoroughly understood by members of the public or private fire service. The knowledge of fire protection personnel is tested when it becomes necessary to establish the following:

- The quantity and pressure of water needed to provide adequate fire protection

- The ability of existing water supply systems to provide fire protection needs

- The adequacy of a water supply in conjunction with a particular piping system to provide effective automatic sprinkler protection

- Alternatives for supporting deficient water supply systems

A training manual such as this one cannot begin to cover the nuances of every fire district. That information can only be gained by continually studying the water supply in a given jurisdiction. This chapter provides information on water supply requirements for automatic extinguishing systems within individual properties and fire department operations in support of those systems. Areas addressed include storage and distribution systems for automatic sprinklers, fire pumps, special-hazard water spray systems, wet and dry standpipes, and other water-based extinguishing systems commonly found on private industrial and commercial properties. It does not provide information on planning for public water supply systems.

NOTE: Please note the variety of IFSTA and FPP publications and videos for more information on water and water supply:

- FPP *Fire Protection Hydraulics and Water Supply Analysis* manual
- IFSTA **Pumping Apparatus Driver/Operator Handbook**
- IFSTA **Essentials of Fire Fighting** manual

Characteristics of Water

Water is a compound of hydrogen and oxygen formed when two parts hydrogen (H) combine with one part oxygen (O) to form H_2O. Between 32°F and 212°F (0°C and 100°C), water exists in a liquid state **(Figure 3.1, p. 88)**. Below 32°F (0°C) (the freezing point of water), water experiences a phase change to a solid state of matter called ice. Above 212°F (100°C) (the boiling point of water), it vaporizes into water vapor or steam. Water cannot be seen in vapor form; it only becomes visible as it rises away from the surface of the liquid water and begins to condense.

For all practical purposes, water is considered to be incompressible, and its weight varies at different temperatures. Water's density, or its weight per unit of volume, is measured in pounds per cubic foot (kg/L). Water is heaviest (approximately 62.4 lb/ft^3 [1 kg/L]) and has its highest density close to its freezing point. Water is lightest (approximately 60 lb/ft^3 [0.96 kg/L]) and has its lowest density close to its boiling point. For fire protection purposes, ordinary fresh water is generally considered to weigh 62.5 lb/ft^3 or 8.33 lb/gal (1 kg/L).

Extinguishing Properties of Water

Water has the ability to extinguish fire in several ways. The primary way in which water extinguishes fire is by cooling, namely absorbing heat from the fire. Another way is by smothering, which occurs

Solid
Ice

32°F (0°C)

Liquid
Water

32° to 212°F (0°C to 100°C)

Gaseous
Invisible
Water Vapor

212°F (100°C)

Increasing Temperature

Figure 3.1 Example of water in three states: solid, liquid, and gas. The transition from one state to another is due to the increase or decrease in the temperature of the water.

when oxygen is excluded. This works especially well on the surface of heavy flammable liquids. Smothering also occurs to some extent when water converts to steam in a confined space.

As an extinguishing agent, water is affected by two natural laws of physics: The Law of Specific Heat and The Law of Latent Heat of Vaporization. These laws are vitally important when considering the heat-absorbing ability of water. The amount of heat absorbed by water is also affected by the amount of surface area of the water exposed to the heat. Another important consideration is specific gravity.

The Law of Specific Heat

The term *specific heat* is a measure of the heat-absorbing capacity of a substance. Water is not only noncombustible, but it can also absorb large amounts of heat. Amounts of heat transfer are measured in the customary system as British thermal units (BTUs) or in the SI system as joules (J) (1 Btu = 1.055 kJ). A BTU is the amount of heat required to raise the temperature of 1 pound of water 1 degree F. The joule, also a unit of work, has taken the place of the calorie in the SI (International System of Units) heat measurement (1 calorie = 4.19 joules).

The specific heat of any substance is the ratio between the amount of heat needed to raise the temperature of a specified quantity of a material and the amount of heat needed to raise the temperature of an identical quantity of water by the same number of degrees. The specific heat of different substances varies. **Table 3.1** shows some fire extinguishing agents and the specific heat comparison (by weight) with water.

Using Table 3.1, divide the specific heat of water (1.00) by the specific heat of carbon dioxide gas (0.19). Note that it takes more than five times the amount of heat to raise the temperature of 1 pound of water 1 degree F than it takes to raise the tem-

Table 3.1 Specific Heat of Extinguishing Agents	
Agent	**Specific Heat**
Water	1.00
Calcium chloride solution	0.70
Carbon dioxide (solid)	0.12
Carbon dioxide (gas)	0.19
Sodium bicarbonate	0.22

perature of the same amount of carbon dioxide gas (**Figure 3.2**). Another way of stating this is that water absorbs five times as much heat as does an equal amount of carbon dioxide. A comparison of the materials listed shows that water is clearly the best material for absorbing heat.

Figure 3.2 It takes five times as much heat to raise the temperature of 1 pound of water 1 degree F as it does an equal amount of carbon dioxide.

The Law of Latent Heat of Vaporization

The *latent heat of vaporization* is the quantity of heat absorbed by a substance when it changes from a liquid to a vapor. The temperature at which a liquid absorbs enough heat to change to vapor is known as its *boiling point*. At sea level, water begins to boil or vaporize at 212°F (100°C). Vaporization, however, does not completely occur the instant water reaches the boiling point. Each pound of water requires approximately 970 Btu (1 023 kJ) of additional heat to completely convert into steam (**Figures 3.3 a and b**).

The latent heat of vaporization is significant in fighting fire because the temperature of the water is not increased beyond 212°F during the absorption of the 970 Btu for every pound of water. Suppose we have 1 gallon (U.S.) of water at 60°F. Each gallon of water weighs 8.33 pounds. It requires 152 Btu (212 – 60 = 152) to raise each pound of water to 212°F. Thus, 1 gallon of water absorbs 1,266 Btu (152 Btu/lb × 8.33 lb) getting to 212°F. Because the conversion to steam requires another 970 Btu per pound, an additional 8,080 Btu (970 Btu/pound × 8.33 lb) will be absorbed through this process. This means that 1 gallon of water will absorb 9,346 Btu

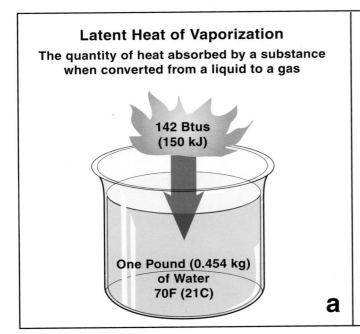

Figure 3.3a To bring water from 70°F (21°C) to its boiling point (212°F [100°C]), 142 Btus (150 kJ) are needed.

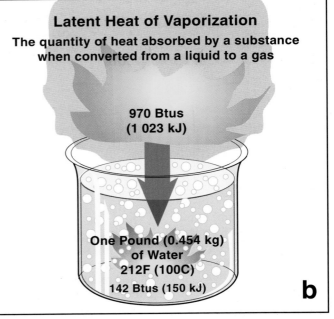

Figure 3.3b When water has reached its boiling point, 970 additional Btus (1 023 kJ) are required to turn it into steam. This high heat of vaporization makes water a good extinguishing agent.

(1,266 + 8,080) of heat if all the water is converted to steam. If water from a 100 gpm (400 L/min) fog nozzle (at 60°F) is projected into a highly heated area, it can absorb approximately 934,600 Btu of heat per minute if all of the water is converted to steam.

The amount of heat a combustible object can produce depends upon the material from which it is composed. The rate at which the object gives off heat depends upon such factors as its physical form, amount of surface exposed, and air or oxygen supply.

Surface Area of Water

The speed with which water absorbs heat increases in proportion to the water surface exposed to the heat. For example, if a 1-inch (25.4 mm) cube of ice is dropped into a glass of water, it will take quite awhile for the ice cube to absorb its capacity of heat (melt). This is because only 6 square inches (3 870 mm²) of the ice are exposed to the water (**Figure 3.4**). If the same cube of ice is divided into 1/8-inch (3 mm) cubes and these cubes are dropped into the water, 48 square inches (30 970 mm²) of the ice are exposed to the water. Although the smaller cubes equal the same mass of ice as the larger cube, the smaller cubes melt faster. That is why crushed ice

melts in a drink faster than ice cubes. This principle also applies to water in a liquid state. If water is divided into many drops, the rate of heat absorption increases hundreds of times.

Another characteristic of water that aids in fire fighting is its expansion capability when converted to steam. This expansion helps cool the fire by driving heat and smoke from the area. The amount of expansion that occurs varies with the temperatures of the fire area. At 212°F (100°C), water expands approximately 1,700 times its original volume (**Figure 3.5**).

To illustrate steam expansion, consider a nozzle discharging 150 gallons (600 L) of water fog every minute into an area heated to approximately 500°F (260°C), causing the water fog to convert to steam (**Figure 3.6**). During one minute of operation, 20 cubic feet (0.57 m³) of water is discharged and vaporized. This 20 cubic feet (0.57 m³) of water expands to approximately 48,000 cubic feet (1 359 m³) of steam. This is enough steam to fill a room approximately 10 feet (3 m) high, 50 feet (15 m) wide, and 96 feet (29 m) long. In hotter atmospheres, steam expands to even greater volumes.

Steam expansion is not gradual, but rapid. If a room is already filled with smoke and gases, the steam generated displaces these gases when adequate ventilation openings are provided (**Figure 3.7**). As the room cools, the steam condenses and allows the room to refill with cooler air. The use of a fog stream in a direct or combination fire attack requires that adequate ventilation be provided ahead of the hoseline. Otherwise, there is a high possibility of steam or even fire rolling back over and around the hose team, and the potential for

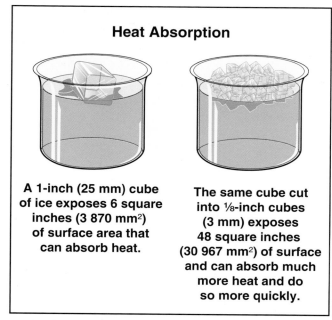

Heat Absorption

A 1-inch (25 mm) cube of ice exposes 6 square inches (3 870 mm²) of surface area that can absorb heat.

The same cube cut into ⅛-inch cubes (3 mm) exposes 48 square inches (30 967 mm²) of surface and can absorb much more heat and do so more quickly.

Figure 3.4 Dividing water into smaller particles increases its rate of heat absorption.

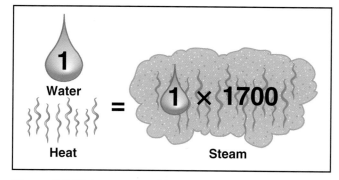

Figure 3.5 Water converted into steam expands 1,700 times. This expansion process absorbs heat and forces hot air and combustion gases out of a confined space.

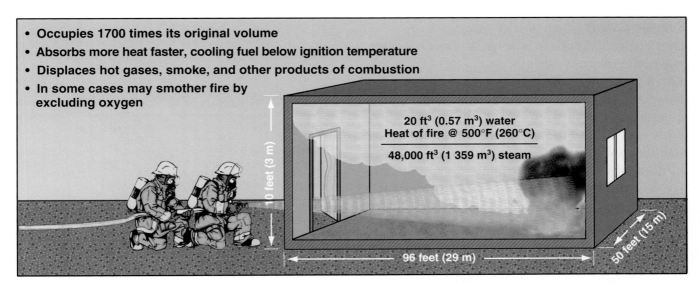

- **Occupies 1700 times its original volume**
- **Absorbs more heat faster, cooling fuel below ignition temperature**
- **Displaces hot gases, smoke, and other products of combustion**
- **In some cases may smother fire by excluding oxygen**

20 ft³ (0.57 m³) water
Heat of fire @ 500°F (260°C)

48,000 ft³ (1 359 m³) steam

10 feet (3 m)

96 feet (29 m)

50 feet (15 m)

Figure 3.6 A nozzle discharging 1,650 gallons (600 L) of water fog for one minute generates enough steam to fill a room approximately 10 feet (3 m) high, 50 feet (15 m) wide, and 96 feet (29 m) long.

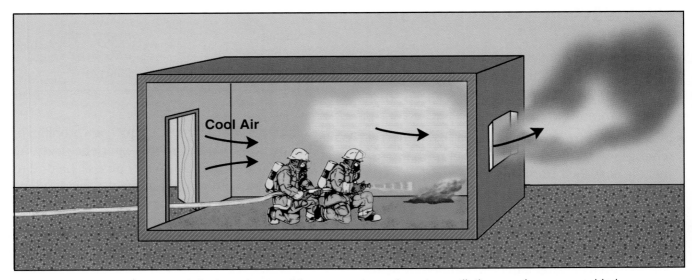

Cool Air

Figure 3.7 The steam generated displaces heat and fire gases when adequate ventilation openings are provided.

burns or steam burns is great. There are some observable results of the proper application of a water fire stream into a room: Fire is extinguished or reduced in size, visibility may be maintained, and room temperature is reduced.

Water can also smother fire when it floats on liquids that have a higher specific gravity than water, such as carbon disulfide. If the material is water soluble, such as alcohol, the smothering action is not likely to be effective. Water may also smother fire by forming an emulsion over the surface of certain combustible liquids. When a spray of water agitates the surface of these liquids, the agitation causes the water to be temporarily suspended in emulsion bubbles on the surface. The emulsion bubbles then smother the fire. This action can only take place when the combustible liquid has sufficient viscosity. *Viscosity* is the tendency of a liquid to possess internal resistance to flow. For example, water has low viscosity; molasses has high viscosity. A heavy fuel oil, such as No. 6 grade, retains an emulsified surface longer than a lighter grade such as No. 2 home heating fuel or diesel fuel. The water in the emulsion absorbs heat from the oil adjacent to it, reduces the oil temperature, and decreases the amount of combustible vapors that are emitted.

Specific Gravity

The density of liquids in relation to water is known as *specific gravity*. Water is given a value of 1. Liquids with a specific gravity less than 1 are lighter than water and therefore float on water. Those with a specific gravity greater than 1 are heavier than water and sink to the bottom. If the other liquid also has a specific gravity of 1, it mixes evenly with water. Most flammable liquids have a specific gravity of less than 1 (**see Table 3.2**). Therefore, if a firefighter flows water on a flammable liquid fire improperly, the whole fire can just float away on the water and ignite everything in its path. The use of foam can control this situation because it floats on the surface of the flammable liquid and smothers the fire.

Advantages and Disadvantages of Water

Water has a number of characteristics that make it an excellent extinguishing agent:

- Water has a greater heat-absorbing capacity than other common extinguishing agents.

- A relatively large amount of heat is required to change water into steam. This means that more heat is absorbed from the fire. Note that water converted into steam occupies 1,700 times its original volume.

- The greater the surface area of water exposed, the more rapidly heat is absorbed. The exposed surface area of water can be expanded by using fog streams or by deflecting solid streams off objects.

- Water is plentiful and readily available in most jurisdictions.

The following disadvantages are due to some additional properties that water possesses:

- Water has a high surface tension and does not readily soak into dense materials. However, when wetting agents are mixed with water, the water's surface tension is reduced and its penetrating ability is increased.

- Water may be reactive with certain fuels such as combustible metals.

- Water has low levels of opacity and reflectivity that allow radiant heat to easily pass through it.

Table 3.2 Specific Gravity of Common Substances	
Substance	**Specific Gravity at 68°F**
Water	0.998
Commercial Solvent	0.717
Carbon Tetrachloride	1.582
Medium Lubrication Oil	0.891
Medium Fuel Oil	0.854
Heavy Fuel Oil	0.908
Regular Gasoline	0.724
Turpentine	0.862
Ethyl Fuel	0.789
Benzene	0.879
Glycerin	1.262
Light Machinery Oil	0.907
Air	0.0012
Ammonia	0.0007
Carbon Dioxide	0.0018
Methane	0.0017

- Water freezes at 32°F (0°C), which is a problem in jurisdictions that frequently experience freezing temperatures. Freezing water poses a hazard to firefighters by coating equipment, roofs, ladders, and other surfaces (**Figure 3.8**). In addition, ice forming in and on equipment may cause it to malfunction.

- Water readily conducts electricity, which can be hazardous to firefighters working around energized electrical equipment.

Understanding Water Pressure

The word *pressure* has a variety of meanings. Ordinarily, one thinks of pressure as force exerted on one substance by another. In this manual, however, pressure is defined as force per unit area in a liquid or gas. Pressure may be expressed in pounds per square foot (psf), pounds per square inch (psi), or kilopascals (kPa).

Pressure can easily be confused with force. *Force* is a simple measure of weight and is usually expressed in pounds or kilograms. This measurement is directly related to the force of gravity, which is the amount of attraction the earth has for all bodies.

Figure 3.8 A significant disadvantage of water is the hazard it poses to firefighters and equipment during freezing conditions. *Courtesy of District Chief Chris E. Mickal.*

If several objects of the same size and weight are placed on a flat surface, they each exert the same force on that surface.

For example, three square containers of equal size (1 × 1 × 1 foot [0.3 m by 0.3 m by 0.3 m]) containing 1 cubic foot (0.028 m³) of water and weighing 62.5 pounds (28 kg) each are placed next to each other **(Figure 3.9)**. Each container exerts a force of about 62.5 psf (about 306 kg/m²) with a total of about 187.5 pounds (85 kg) of force over a 3-square-foot (0.3 m²) area.

If the containers are stacked on top of each other, the total force exerted — 187.5 pounds or 85 kg — remains the same, but the area of contact is reduced to 1 square foot (0.1 m²)), the pressure then becomes 187.5 psf (about 919 kg/m²) **(Figure 3.10, p. 94)**.

To understand how force is determined, it is necessary to know the weight of water and the height that a column of water occupies. The weight of 1 cubic foot of water is approximately 62.5 pounds. Because 1 square foot contains 144 square inches, the weight of water in a 1-square-inch column of water 1 foot high equals 62.5 pounds divided by 144 square inches or 0.434 pounds. A 1-square-inch column of water 1 foot high therefore exerts a pressure at its base of 0.434 psi **(Figure 3.11, p. 94)**. The

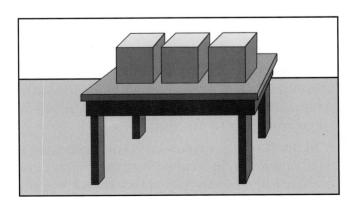

Figure 3.9 When placed next to each other, each container exerts a pressure of 62.5 pounds per square foot (306 kg/m²).

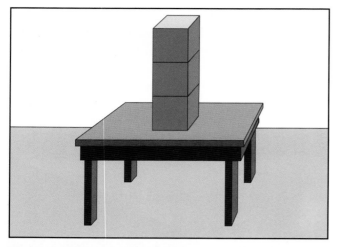

Figure 3.10 If the same containers are stacked, they exert a pressure of 187.5 pounds per square foot (919 kg/m²).

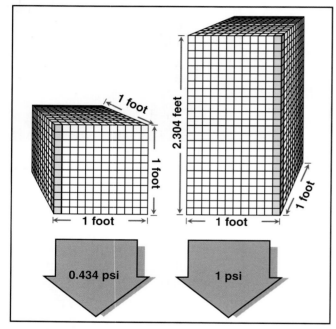

Figure 3.11 These examples clearly illustrate the relationship between height and pressure in the U.S. system of measurement.

height required for a 1-square-inch column of water to produce 1 psi at its base equals 1 foot divided by 0.434 psi/ft or 2.304 feet; therefore, 2.304 feet of water column exerts a pressure of 1 psi at its base.

In the SI system of measurement, a cube that is 0.1 m × 0.1 m × 0.1 m (a cubic decimeter) holds 1 liter of water. The weight of 1 liter of water is 1 kilogram. The cube of water exerts 1 kPa (1 kg) of pressure at the bottom of the cube. One cubic meter of water

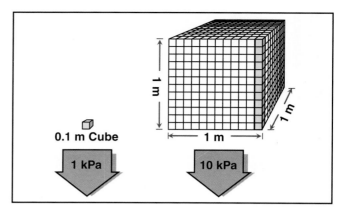

Figure 3.12 An illustration of the relationship between height and pressure in the metric system of measurement.

holds 1 000 liters of water and weighs 1 000 kg. Because the cubic meter of water is comprised of 100 columns of water, each 10 decimeters tall, each column exerts 10 kPa at its base (**Figure 3.12**).

Principles of Pressure

The speed at which a fluid travels through hose or pipe is developed by pressure upon that fluid. The speed at which this fluid travels is often referred to as *velocity head pressure*. The speed with which a fluid travels through hose or pipe is a result of the pressure exerted on the fluid at its source. It is important to identify the type of pressure because the word pressure in connection with fluids has a very broad meaning. There are six basic principles that determine the action of pressure upon fluids, and it is very important that fire service personnel clearly understand the various sources of pressure.

NOTE: The density of water is affected by temperature. When making hydraulic calculations where absolute accuracy is necessary, the temperature of water must be considered.

First Principle

Fluid pressure is perpendicular to any surface on which it acts. This principle is illustrated by observing a vessel that has flat sides and contains water (**Figure 3.13**). The pressure exerted by the weight of the water is perpendicular to the walls of the container. If this pressure is exerted in any other direction, as indicated by the slanting arrows, the water would start moving downward along the sides and rising in the center.

Second Principle

Fluid pressure at a point in a fluid at rest is the same intensity in all directions. To put it another way, fluid pressure at a point in a fluid at rest has no direction (**Figure 3.14**).

Third Principle

Pressure applied to a confined fluid from without is transmitted equally in all directions (**Figure 3.15**). This principle is illustrated by viewing a hollow sphere to which a water pump is attached. A series of gauges is set into the sphere around its circumference. When the sphere is filled with water and pressure is applied by the pump, all gauges will register the same pressure. This is true if they are on the same grade line with no change in elevation.

Fourth Principle

The pressure of a liquid in an open vessel is proportional to its depth (**Figure 3.16**). This principle is illustrated by observing three vertical containers, each 1 square inch (645 mm²) in a cross-sectional area. The depth of the water is 1 foot (0.3 m) in the first container, 2 feet (0.6 m) in the second, and 3 feet (0.9 m) in the third container. The pressure at the bottom of the second container is twice that of the first, and the pressure at the bottom of

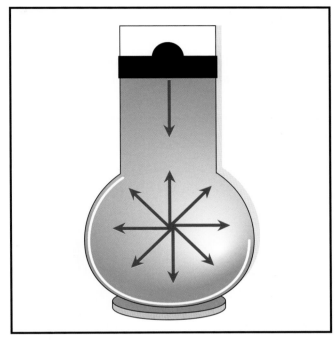

Figure 3.15 Pressure that is transmitted to a confined fluid from without is transmitted equally in all directions.

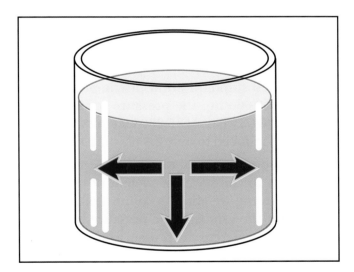

Figure 3.13 The pressure exerted by the weight of the fluid is perpendicular to the walls of the container.

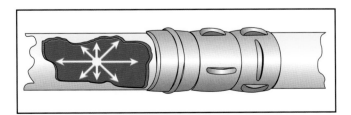

Figure 3.14 When a fluid is at rest, fluid intensity is the same in all directions.

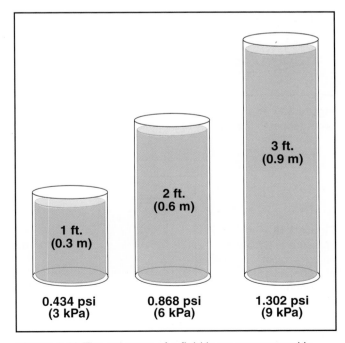

3 ft. (0.9 m)

2 ft. (0.6 m)

1 ft. (0.3 m)

0.434 psi (3 kPa)   0.868 psi (6 kPa)   1.302 psi (9 kPa)

Figure 3.16 The pressure of a fluid in an open vessel is proportional to its depth.

the third container is three times that of the first. Thus, the pressure of a liquid in an open container is proportional to its depth.

Fifth Principle

The pressure of a liquid in an open vessel is proportional to the density of the liquid. This principle is illustrated by observing two containers. One container holds mercury 1 inch (25 mm) deep, the other holds water 13.55 inches (344 mm) deep, yet the pressure at the bottom of each container is approximately the same (**Figure 3.17**). Thus, mercury

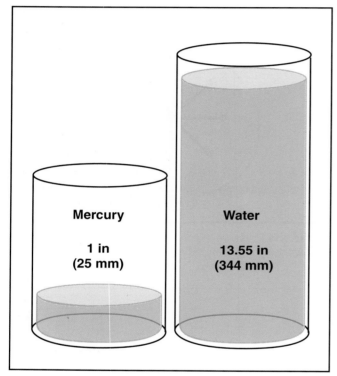

Figure 3.17 The pressure of a fluid in an open vessel is proportional to the density of the fluid.

is 13.55 times denser than water. Therefore, the pressure of a liquid in an open vessel is proportional to the density of the liquid.

Sixth Principle

The pressure of a liquid on the bottom of a vessel is independent of the shape of the vessel. This principle is illustrated by showing water in several different-shaped containers (**Figure 3.18**). The pressure is the same in each vessel regardless of the shape of the opening.

Types of Pressure

There are a number of terms used for different types of pressure that are encountered in water supply systems and in the fire service. The driver/operator should be acquainted with each of these terms so that they can be used in their proper context.

Atmospheric Pressure

Atmospheric pressure is the pressure exerted on the earth by the atmosphere itself. The atmosphere surrounding the earth has depth and density and exerts pressure upon everything on earth. Atmospheric pressure is greatest at low altitudes and least at very high altitudes. At sea level, the atmosphere exerts a pressure of 14.7 psi (101 kPa), which is considered standard atmospheric pressure. At 5,000 feet (1 666 m), the pressure is reduced to 12.228 psi (84.3 kPa). At 15,000 feet (5 000 m), the pressure is reduced to only 8.3 psi (57.2 kPa).

A common method of measuring atmospheric pressure is to compare the weight of the atmosphere with the weight of a column of mercury: the greater the atmospheric pressure, the taller the column of mercury. A pressure of 1 psi (7 kPa)

Figure 3.18 The pressure of a fluid on the bottom of a container is independent of the container's shape.

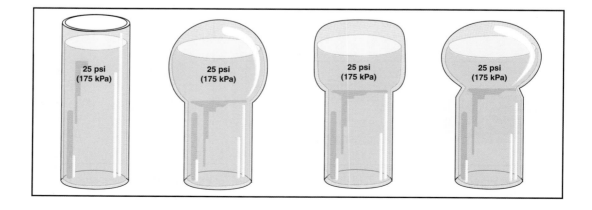

makes the column of mercury about 2.04 inches (52 mm) tall. At sea level, then, the column of mercury is 2.04 × 14.7, or 29.9 inches (759 mm) tall **(Figure 3.19)**.

The readings of most pressure gauges are psi (or kPa) above the existing atmospheric pressure. For example, a gauge reading 10 psi (70 kPa) at sea level is actually indicating a total pressure of 24.7 psi (170 kPa) (14.7 + 10 [100 + 70 kPa]). Engineers distinguish between such a gauge reading and total atmospheric pressure by writing *psig*, which means "pounds per square inch gauge." The notation for actual atmospheric pressure is *psia*, meaning "pounds per square inch absolute" (the psi above a perfect vacuum, absolute zero). Any pressure less than atmospheric pressure is called *vacuum*. Absolute zero pressure is called a *perfect vacuum*. When a gauge reads -5 psig (-35 kPa), it is actually reading 5 psi (35 kPa) less than the existing atmospheric pressure (at sea level, 14.7 minus 5, or 9.7 psia [100 minus 35, or 65 kPa]).

NOTE: Throughout this manual, psi means psig.

Head Pressure

The term *head* in the fire service is another way of expressing pressure. It refers to the height that a pressure can lift a column of liquid. The height of a water supply above the discharge orifice is called the *elevation head*. In **Figure 3.20**, the water supply is 100 feet (30 m) above the hydrant discharge opening. This is referred to as 100 feet (30 m) of head. To convert head in feet (meters) to head pressure, divide the number of feet by 2.304 (for metric, divide the number of meters by 0.1). The result is the number of feet (meters) that 1 psi (7 kPa) raises a column of water. The water source in Figure 3.20 has a head pressure of 43.4 psi (300 kPa) **(Table 3.3, p. 98)**.

Static Pressure

The water flow definition of *static pressure* is stored potential energy available to force water through pipe, fittings, fire hose, and adapters. *Static* means at rest or without motion. Pressure on water may be produced by an elevated water supply, by atmospheric pressure, or by a pump. If the water is not moving, the pressure exerted is static. A truly static pressure is seldom found in municipal water systems because there is always some flow in the pipes due to normal domestic or industrial needs. Nevertheless, the pressure in a water system before water flows from a hydrant is considered static pressure **(Figure 3.21, p. 99)**.

Normal Operating Pressure

Normal operating pressure is that pressure found in a water distribution system during normal consumption demands. As soon as water starts to

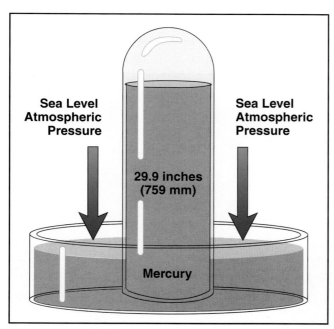

Figure 3.19 A pressure of 14.7 psi (101 kPa) causes the mercury column in this barometer to rise 29.9 inches (759 mm).

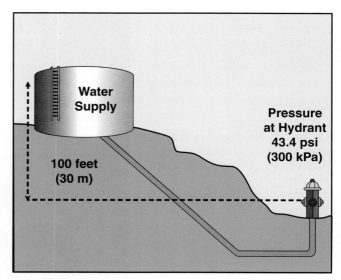

Figure 3.20 The head in this illustration is 100 feet (30 m). The head pressure is 43.4 psi (300 kPa).

Table 3.3
Head in Feet (Meters) and Head Pressure

US Feet of Head	Pounds per Square Inch	Pounds per Square Inch	Feet of Head	METRIC Meters of Head	kPa	kPa	Meters of Head
5	2.17	5	11.50	1	10	5	.5
10	4.33	10	23.00	2	20	10	1
15	6.5	15	34.60	3	30	15	1.5
20	8.66	20	46.20	4	40	20	2
25	10.83	25	57.70	5	50	25	2.5
30	12.99	30	69.30	6	60	30	3
35	15.16	35	80.80	7	70	35	3.5
40	17.32	40	92.30	8	80	40	4
50	21.65	50	115.40	9	90	50	5
60	26.09	60	138.50	10	100	60	6
70	30.30	70	161.60	15	150	70	7
80	34.60	80	184.70	20	200	80	8
90	39.00	90	207.80	25	250	90	9
100	43.30	100	230.90	30	300	100	10
120	52.00	120	277.00	40	400	200	20
140	60.60	140	323.20	50	500	300	30
160	69.20	160	369.40	60	600	400	40
200	86.60	180	415.60	70	700	500	50
300	129.90	200	461.70	80	800	600	60
400	173.20	250	577.20	90	900	700	70
500	216.50	275	643.00	100	1 000	800	80
600	259.80	300	692.70	200	2 000	900	90
800	346.40	350	808.10	300	3 000	1 000	100
1,000	433.00	500	1,154.50				

Figure 3.21 The pressure in a water system before water flows from a hydrant is considered static pressure.

flow through a distribution system, static pressure no longer exists. The demands for water consumption fluctuate continuously, causing water flow to increase or decrease in the system. The difference between **true** static pressure and normal operating pressure is the friction caused by water flowing through the various pipes, valves, and fittings in the system.

Residual Pressure

Residual pressure is that part of the total available pressure not used to overcome friction loss or gravity while forcing water through pipe, fittings, fire hose, and adapters. *Residual* means a remainder or that which is left. In a water distribution system, residual pressure varies according to the amount of water flowing, the size of the pipe, and any other restrictions that are present. For example, during a fire flow test, residual represents the pressure left in a distribution system within the vicinity of one or more flowing hydrants. In a water distribution system, residual pressure varies according to the amount of water flowing from one or more hydrants, water consumption demands, and the size of the pipe. It is important to remember that residual pressure is measured at the location where a pressure reading is taken, not at point of flow.

Flow Pressure (Velocity Pressure)

Flow pressure is that forward velocity pressure while water is flowing (**Figure 3.22**). Because a stream of water emerging from a discharge opening is not encased within a tube, it exerts forward pressure but not sideways pressure. The forward velocity of flow pressure can be measured by using a pitot tube and gauge. If the size of the opening is known, a firefighter can use the measurement of flow pressure to calculate the quantity of water flowing in gpm or L/min.

Pressure Loss and Gain: Elevation and Altitude

Although the words elevation and altitude are often used interchangeably, the fire service makes a distinction between the two. *Elevation* refers to the center line of the pump or the bottom of a static water supply source above or below ground level. *Altitude* is the position of an object above or

Figure 3.22 The flow pressure is usually measured by inserting a pitot tube and gauge in the stream flowing from an open hydrant discharge.

below sea level. Both are important in producing effective fire streams.

When a nozzle is above the pump, there is a pressure loss. When the nozzle is below the pump, there is a pressure gain. These losses and gains occur because of gravity. Both pressure loss and pressure gain are referred to as *elevation head pressure*.

Altitude affects the production of fire streams because atmospheric pressure drops as height above sea level increases. This pressure drop is of little consequence up to about 2,000 feet (600 m). Above this height, though, the lessened atmospheric pressure can be of concern. At high altitudes, fire department pumpers must work harder to produce the pressures required for effective fire streams while drafting. They must work harder because lower atmospheric pressure reduces the pumper's effective lift when drafting. Above sea level, atmospheric pressure decreases approximately 0.5 psi (3.5 kPa) for every 1,000 feet (300 m).

Friction Loss

The fire service definition of *friction loss* is that part of the total pressure lost while forcing water through pipe, fittings, fire hose, and adapters. The common term for pressure loss due to friction is simply friction loss. It is important for fire service professionals to understand the principles of friction loss in order to be able to effectively assess water flow. In a fire hose, friction loss is caused by the following:

- Movement of water molecules against each other
- Linings in fire hose
- Couplings
- Sharp bends
- Change in hose size or orifice by adapters
- Improper gasket size

Anything that affects movement of water may cause additional friction loss. Good-quality fire hose has a smoother inner surface and causes less friction loss than lower-quality hose. The friction loss in old hose may be as much as 50 percent greater than that in new hose.

The principles of friction loss in piping systems are the same as in fire hose. Friction loss is caused by the following:

- Movement of water molecules against each other
- Inside surface of the piping
- Pipe fittings
- Bends
- Control valves

The rougher the inner surface of the pipe (commonly referred to as the *coefficient of friction*), the more friction loss that occurs. Friction loss can be measured by inserting in-line gauges in a hose or pipe. The difference in the residual pressures between gauges when water is flowing is the friction loss. The difference in pressure in a fire hose between a nozzle and a pumper is a good example of friction loss.

Principles of Friction Loss

There are four basic principles that govern friction loss in fire hose and pipes. These principles are discussed in the sections that follow.

First Principle

If all other conditions are the same, friction loss varies proportionately with the length of the hose or pipe. This principle can be illustrated by one hose that is 100 feet (30 m) long and another hose that is 200 feet (60 m) long (**Figure 3.23**). A constant flow of 200 gpm (800 L/min) is maintained in each hose. The 100-foot (30 m) hose has a friction loss of

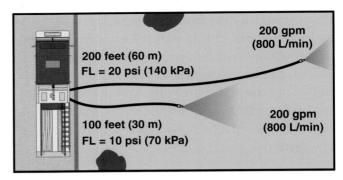

Figure 3.23 With all other variables the same, friction loss varies directly with the length of the hose.

10 psi (70 kPa). The 200-foot (60 m) hose has twice as much friction loss, or 20 psi (140 kPa).

Second Principle

When hoses are the same size, friction loss varies approximately with the square of the increase in the velocity of the flow (**Figure 3.24**). This principle points out that friction loss develops much faster than the change in velocity. (Remember that velocity is proportional to flow.) For example, a length of 3-inch (77 mm) hose flowing 200 gpm (800 L/min) has a friction loss of 3.2 psi (22.4 kPa). As the flow doubles from 200 to 400 gpm (800 L/min to 1 600 L/min), the friction loss increases four times ($2^2 = 4$) to 12.8 psi (89.6 kPa). When the original flow is tripled from 200 to 600 gpm (800 L/min to 2 400 L/min), friction loss increases nine times ($3^2 = 9$) to 28.8 psi (201.6 kPa).

Third Principle

For the same discharge, friction loss varies inversely as the fifth power of the diameter of the hose. This principle readily proves the advantage of larger size hose and can be illustrated by one hose that is 2½ inches (65 mm) in diameter and another that is 3 inches (77 mm) in diameter. The friction loss in the 3-inch (77 mm) hose is:

$$\frac{(2\ \tfrac{1}{2})^5}{3^5} = \frac{98}{243} = 0.4 \text{ that of the 2½-inch hose}$$

$$\frac{(65)^5}{(77)^5} = \frac{1\ 160\ 290\ 625}{2\ 706\ 784\ 157} = 0.4 \text{ that of the 65 mm hose}$$

Fourth Principle

For a given flow velocity, friction loss is approximately the same regardless of the pressure on the water. This principle explains why friction loss is

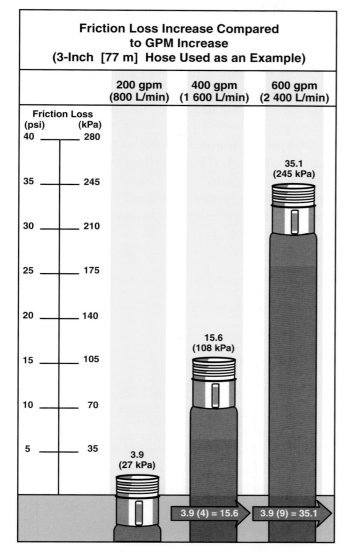

Friction Loss Increase Compared to GPM Increase (3-Inch [77 m] Hose Used as an Example)

	200 gpm (800 L/min)	400 gpm (1 600 L/min)	600 gpm (2 400 L/min)

Friction Loss (psi) / (kPa)

40 — 280
35 — 245
30 — 210
25 — 175
20 — 140
15 — 105
10 — 70
5 — 35

35.1 (245 kPa)
15.6 (108 kPa)
3.9 (27 kPa)

3.9 (4) = 15.6
3.9 (9) = 35.1

Figure 3.24 When the velocity of the stream is increased, the friction loss increases at an incrementally higher rate.

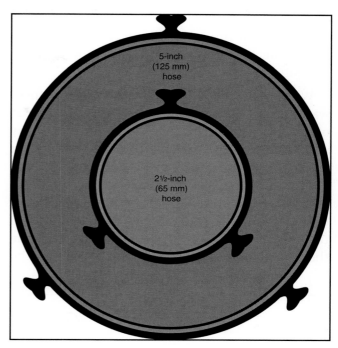

5-inch (125 mm) hose

2½-inch (65 mm) hose

Figure 3.25 Doubling the diameter of the hose increases the area of the hose opening approximately four times.

the same when hoses or pipes at different pressures flow the same amount of water. For example, if 100 gpm (400 L/min) passes through a 3-inch (77 mm) hose within a certain time, the water must travel at a specified velocity (feet per second [meters per second]). For the same rate of flow to pass through a 1½-inch (38 mm) hose, the velocity must be greatly increased. Four 1½-inch (38 mm) hoses are needed to flow 100 gpm (400 L/min) at the same velocity required for a single 3-inch (77 mm) hose **(Figure 3.25)**.

While pipe sizes are fixed, some brands of fire hose tend to expand to a larger inside diameter under higher pressures than other brands. Even though both brands of hose may be marketed as

1¾-inch (45 mm) hose, one brand may expand to nearly 2 inches (50 mm) when charged. Keep in mind that this tendency to expand decreases the velocity and therefore decreases the friction loss.

Other Factors Affecting Friction Loss

One of the physical properties of water is that it is practically incompressible. This means that the same volume of water supplied into hose or piping under pressure at one end will be discharged at the other end. The diameter of hose/piping determines the velocity for a given volume of water. The smaller the diameter, the greater the velocity needed to deliver the same volume.

Friction loss in a system increases as the length of hose or piping increases. Flow pressure will always be greatest near the supply source and lowest at the farthest point in the system. A condition existing in a practical water system and a fire hose layout is shown in **Figure 3.26, p. 102.** An elevated tank is filled with water to a height of 150 feet (50 m). The pipe connections to the fire hydrant are at the bottom of the tank. From the hydrant, 300 feet (90 m) of 2½-inch (65 mm) hose is laid along the street with a valve on the end. Imagine a glass tube connected to the hose every 100 feet (30 m) standing upright 150 feet (45 m) to the same height as the elevated tank. With the valve closed, the water in all the tubes would stand at Line A, which is the same level as the water in the tank. This line indicates the static pressure.

When the valve on the nozzle end is opened, water flows moderately at a low pressure. If the opening is made directly at the hydrant, the flow will be much greater at a higher pressure. In other words, the flow pressure is not as great at the end of the hose as it is at the hydrant. Instead of the water being up to the level of the standpipe, it is up to Line B. The difference in the water level of the tubes indicates the pressure used to overcome the friction loss in the sections of hose between the

tubes. The loss of pressure at each 100-foot (30 m) interval and the reduced discharge indicate the friction loss in the line.

An open fire hose produces a stream that normally has no use in fire fighting. Some type of nozzle is needed to shape the stream. When a closed nozzle with a 1-inch (25 mm) tip is added to the system, the water level in the glass tubes returns to Line A. When the nozzle is opened, the water level drops to Line C. Notice that Line C in **Figure 3.27** is considerably higher than Line B in **Figure 3.26** and that a fire stream with some reach has been produced.

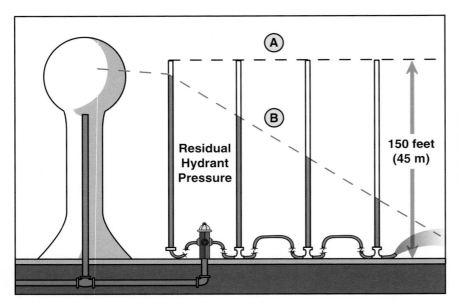

Figure 3.26 Line A indicates the static pressure. The distance between Lines A and B indicates the pressure loss due to friction. Note that there is no nozzle attached to the end of the hose.

Figure 3.27 When a nozzle is attached to the end of the hose, the volume of water flowing is decreased. This results in less friction loss.

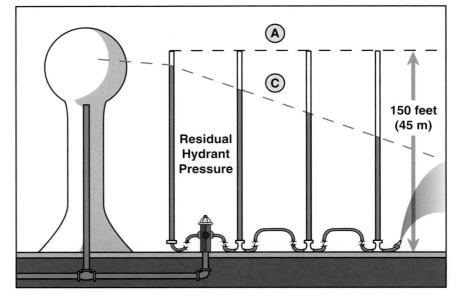

If the 1-inch (25 mm) tip is replaced with a ¾-inch (19 mm) tip, the water level in the glass tubes rises even higher. The velocity increases, but the amount of flow decreases. By decreasing the amount of water flowing, a firefighter reduces the speed of the water in the hose; consequently, there is less friction loss.

Observe the height of the water in the first glass tubes in **Figures 3.26** and **3.27**. The water level in these tubes indicates a good supply of residual pressure left in the water main. Using a pump at this point provides additional force, thus increasing the pressure in the hose. This additional pressure makes it possible to produce effective fire streams. Using a pump also makes it possible to add hose, even to the extent of providing master streams.

It is important to remember that there are practical limits to the velocity or speed at which a stream can travel. If the velocity is increased beyond these limits, the friction becomes so great that the entire stream becomes agitated due to resistance. This agitation causes a degree of turbulence called *critical velocity*.

Reducing Friction Loss

Certain characteristics of hose and piping layouts affect friction loss, including the following:

- Hose/piping length
- Hose/piping diameter

- Sharp bends (kinks) in the hose
- Elbows and other fittings in piping

It is usually possible to minimize sharp bends or kinks in fire hose by using proper hose handling techniques. To reduce friction loss due to hose length or diameter, it is necessary to reduce the length of the hose or increase its diameter. Although the hose must be long enough to reach the needed location, any extra hose should be eliminated to reduce excess friction loss.

Because it is generally safe to increase the flow of water for fire fighting, it is usually acceptable to use larger diameter hose to reduce friction loss.

Water Hammer

Water moving through a pipe or hose has both weight and velocity. The weight of water increases as the pipe or hose size increases. Suddenly stopping water moving through a hose or pipe results in an energy surge being transmitted in the opposite direction, often at many times the original pressure. This surge is referred to as *water hammer*. Water hammer can damage the pump, appliances, hose, or the municipal water system itself (**Figure 3.28**). Always open and close nozzle controls, hydrants, valves, and hose clamps slowly to prevent water hammer. Equip apparatus inlets and remote outlets with pressure relief devices to prevent damage to equipment.

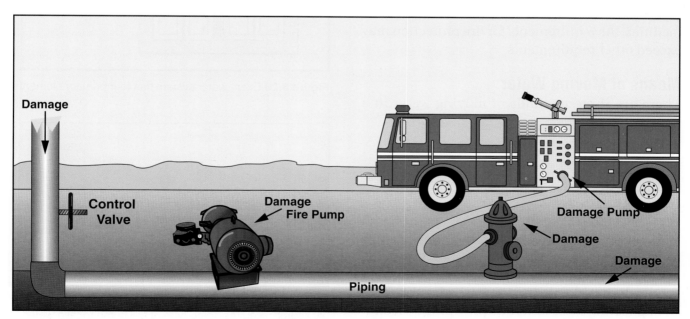

Figure 3.28 Water hammer can damage any part of the water distribution system or any fire apparatus connected to the system.

Principles of Municipal Water Supply Systems

Public and/or private water systems provide the methods for supplying water to areas. As the population in rural areas increases, communities seek to improve water distribution systems from reliable sources. The working parts of a water system are many and varied, but basically the system is composed of the following standard components (**Figure 3.29**):

- Source of water supply
- Means of moving water
- Water processing or treatment facilities
- Water distribution system, including storage

Sources of Water Supply

The primary water supply can be obtained from the public water supply system or from surface water or groundwater. Although most water systems are supplied from only one source, there are instances where both sources are used. Two examples of surface water supply are rivers and lakes (**Figure 3.30**). Groundwater supply can be water wells or water-producing springs.

The amount of water that a facility needs can be determined by an engineering estimate. This estimate is the total amount of water needed for domestic and industrial use and for fire fighting use. In cities, the domestic/industrial requirements far exceed those needed for fire protection. In some facilities, the requirements for fire protection may exceed other requirements.

Means of Moving Water

There are three methods of moving water in a system:

- Direct pumping system
- Gravity system
- Combination system

Direct Pumping System

Direct pumping systems use one or more pumps that take water from the primary source and discharge it into the distribution system. Failures in supply lines and pumps can usually be overcome by duplicating these units and providing a secondary power source (**Figure 3.31**).

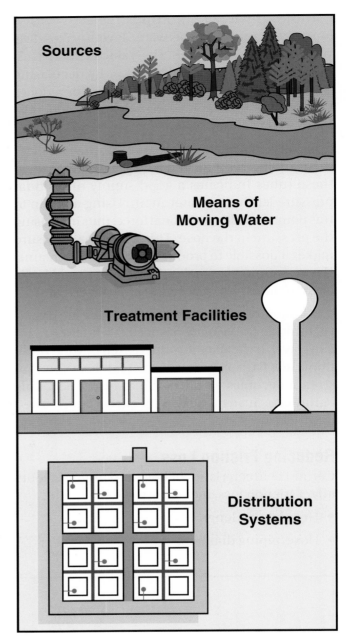

Figure 3.29 Every water system has four primary elements.

Figure 3.30 This lake, which serves as a static water supply source, requires some type of fire pump to take water into the system.

Gravity System

A gravity system uses a primary water source located at a higher elevation than the distribution system **(Figure 3.32)**. The gravity flow from the higher elevation provides the water pressure. This pressure is usually sufficient only when the primary water source is located at least several hundred feet (meters) higher than the highest point in the water distribution system. The most common examples of gravity systems include a mountain reservoir that supplies water to a city below or a system of elevated tanks in a city.

Combination System

Most facilities can use a combination of direct pumping and gravity systems **(Figure 3.33, p. 106)**. In most cases, the gravity flow is supplied by elevated storage tanks. These tanks serve as emergency storage and provide adequate pressure through the use of gravity. When the system pressure is high during periods of low consumption, automatic valves open and allow the elevated storage tanks to fill. When the pressure drops during periods of heavy consumption, the storage containers provide extra water by feeding it back into the distribution system. Providing a good combination system involves reliable, duplicated equipment and proper-sized, strategically located storage containers.

Water stored in elevated reservoirs can also ensure water supply when the system becomes otherwise inoperative. Storage should be sufficient to provide domestic and industrial demands in addition to the demands expected during fire fighting operations. Such storage should also be sufficient to permit making most repairs, alterations, or additions to the system. Location of the storage and the capacity of the mains leading from this storage are also important factors.

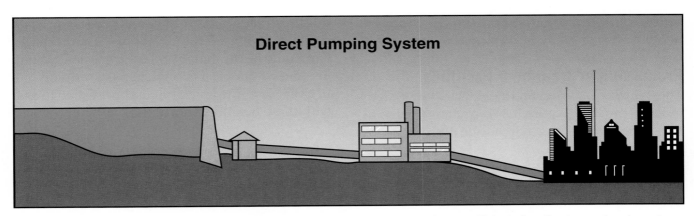

Figure 3.31 A direct pumping system is used when the water source does not have sufficient elevation to create adequate pressure.

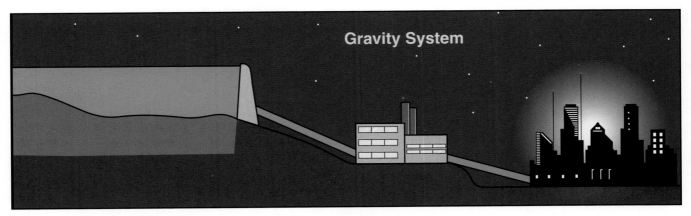

Figure 3.32 A gravity system is used where the water source is elevated.

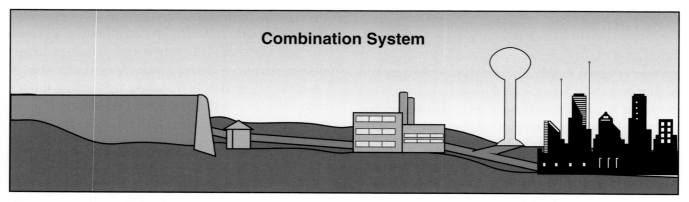

Combination System

Figure 3.33 A combination of direct pumping and gravity is used to allow water storage during low demand. Later, this water can be used when consumption exceeds pump capacity.

Many industries provide their own private systems, such as elevated storage tanks, that are available to the fire department. Water for fire protection may be available to some communities from storage systems, such as cisterns, that are considered a part of the distribution system. The fire department pumper removes the water from these sources by drafting (process of obtaining water from a static source into a pump that is above the source's level) and provides pressure by its pump.

Processing or Treatment Facilities

The treatment of water for the water supply system is a vital process **(Figure 3.34)**. Water is treated to remove contaminants that may be detrimental to the health of those who use or drink it. In addition to removing harmful elements from water, water treatment officials may add fluoride or oxygen. The fire department's main concern regarding treatment facilities is that a maintenance error, natural disaster, loss of power supply, or fire could disable the pumping station(s) or severely hamper the purification process. Any of these situations would drastically reduce the volume and pressure of water available for fire fighting operations. Another problem would be the inability of the treatment system to process water fast enough to meet the demand. In either case, fire officials must have a plan to deal with these potential shortfalls.

Figure 3.34 In most cases, the water treatment facility is located adjacent to the supply source.

Water Distribution System

The distribution system of the overall water supply system receives the water from the pumping station and delivers it throughout the area served. The ability of a water system to deliver an adequate quantity of water relies upon the carrying capacity of the system's network of pipes. When water flows through pipes, its movement causes friction that results in a reduction of pressure. There is much less pressure loss in a water distribution system when fire hydrants are supplied from two or more directions. A fire hydrant that receives water from only one direction is known as a *dead-end hydrant* (**Figure 3.35**). When a fire hydrant receives water from two or more directions, it is said to have *circulating feed* or a *looped line* (**Figure 3.36**). A distribution system that provides circulating feed from several mains constitutes a *grid system*. A grid system should consist of the following components (**Figure 3.37**):

- **Primary feeders** — Large pipes (mains), with relatively widespread spacing, that convey large quantities of water to various points of the system for local distribution to the smaller mains

- **Secondary feeders** — Network of intermediate-sized pipes that reinforce the grid within the various loops of the primary feeder system and aid the concentration of the required fire flow at any point

- **Distributors** — Grid arrangement of smaller mains serving individual fire hydrants and blocks of consumers

To ensure sufficient water, two or more primary feeders should run from the source of supply to the high-risk and industrial districts of the community by separate routes. Similarly, secondary feeders should be arranged in loops as far as possible to give two directions of supply to any point. This

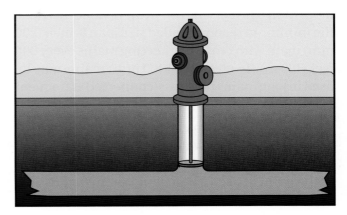

Figure 3.36 Looped hydrants receive water from two directions.

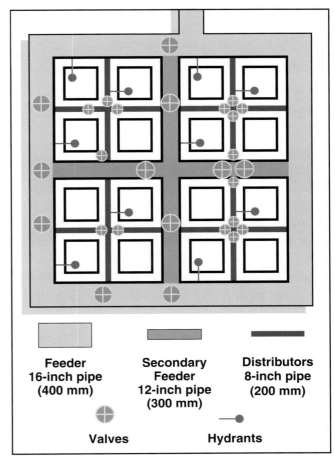

Figure 3.37 A typical grid system of water supply pipes.

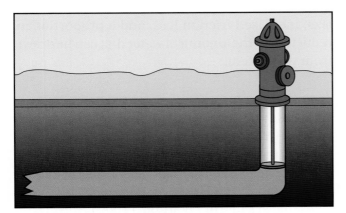

Figure 3.35 Dead-end hydrants do not always supply a reliable flow of water for fire fighting operations.

practice increases the capacity of the supply at any given point and ensures that a break in a feeder main will not completely cut off the water supply.

The recommended size for fire hydrant supply mains is at least 6 inches (150 mm) in diameter. These should be closely gridded by 8-inch (200 mm) cross-connecting mains at intervals of not more than 600 feet (180 m). In high-value districts, the minimum recommended size is an 8-inch (200 mm) main with cross-connecting mains every 600 feet (180 m). Twelve-inch (300 mm) mains may be used in long mains not cross-connected at frequent intervals.

Having a large capacity and reliable supply is of little value if the water cannot be delivered in adequate amounts and with adequate pressure to the point of use. Whether or not this is possible depends in part upon the capacity and pressure rating of the pumps at the treatment plant and pumping stations, as well as the extent of elevated storage. The feature that has the greatest impact is the piping distribution system. The performance of the distribution system is affected by several variables, including the following:

- Piping material
- Pipe diameter
- Piping arrangement

Piping Materials

In the earliest days of water supply systems, wooden piping systems were installed. There is very little of this piping left in the ground. The primary disadvantage of wooden pipe was the limit on interior diameters. It was from the wooden piping that the expression "fireplug" evolved. To access water for fire fighting, holes were drilled in the wood piping and, to stop the flow, plugs were hammered in the holes.

Most communities of any age still have some unlined cast iron piping. This piping corroded and tuberculated easily and was often of small diameter, all resulting in poor flow characteristics. Unlined cast iron is rarely installed today. To improve the performance of the cast-iron pipe, various linings such as cement or asphalt have been used. These linings retard corrosion and improve the hydraulic characteristics of the piping.

A very popular piping material for many years was asbestos cement. Millions of miles of this piping were installed all over the world. This piping would not corrode or tuberculate and maintained excellent hydraulic characteristics for years. However, the problem with the release of asbestos fibers into the air when the pipe was cut or drilled has resulted in its being used only rarely in new installations. Modern piping systems are likely to be of polyvinyl chloride plastic (PVC); however, ductile iron and welded steel piping are also commonly used.

When water mains are installed in unstable or corrosive soils or in difficult access areas, steel or reinforced concrete pipe may be used to provide the strength needed. Some locations that may require extra protection include areas beneath railroad tracks and highways, areas close to heavy industrial machinery, areas prone to earthquakes, or areas of rugged terrain.

The internal surface of the pipe, regardless of the material from which it is made, offers resistance to water flow. Some materials, however, have considerably less resistance to water flow than others. Personnel from the engineering division of the water department should determine the type of pipe best suited for the conditions at hand.

The amount of water able to flow through a pipe and the amount of friction loss created can also be affected by other factors. Frequently, friction loss is increased by encrustation of minerals on the interior surfaces of the pipe. Another problem is sedimentation that settles out of the water. Both of these conditions result in a restriction of the pipe size, increased friction loss, and a proportionate reduction in the amount of water that can be drawn from the system.

Pipe Diameter

Demands for fire protection applications are rarely less than 500 gpm (2 000 L/min). Even when fighting fires in residential areas, fire departments like to have at least 1,000 gpm (4 000 L/min) available. Application of the Hazen-Williams formula indicates that at a rate of 1,000 gpm (4 000 L/min), 20 to 40 psi (140 kPa to 280 kPa) can be lost in each 100 feet (30 m) of 4-inch (100 mm) pipe. For this reason, piping 4 inches (100 mm) and smaller should never be used in water distribution systems that supply fire hydrants. Even piping as small as 6 inches (150 mm) should not be used unless the piping is looped or gridded. In general, water distribution piping should be at least 8 inches (200 mm) in diameter, especially if the piping is dead end. Unfortunately, the water systems in many communities were constructed with little attention to fire protection demands. It is, therefore, common to see 4-inch (100 mm) pipe supplying fire hydrants in the older sections of many cities and towns.

Piping Arrangement

In addition to looping and gridding, piping arrangement is a third feature that affects water distribution system performance. The reduction in pressure loss to friction can be dramatic when dead-end mains are tied together to create a gridded distribution system. For example, if the 8-inch dead-end main is converted to a gridded system as illustrated in **Figure 3.37**, the pressure loss in delivering 2,000 gpm (8 000 L/min) a distance of 1,000 feet (300 m) is reduced from 46 psi to about 7 psi (322 kPa to 49 kPa). A well-gridded system using pipe sizes 8 inch (200 mm) and larger is one of the identifying features of a strong distribution system.

A typical water distribution system also contains pumping stations to boost the pressure to outlying areas or higher elevations in a community. These pumping stations often include ground-level storage tanks from which the pumps take suction. Elevated storage tanks are also common components. Many people mistakenly believe that the pressure within a water system comes from these elevated tanks. They believe this without thinking about how the tanks are filled. The pressure on most water systems comes from the pumps at the treatment plant and pumping stations. This pressure is used to fill the elevated storage tanks. The water in the tanks will not be delivered into the system until the demand upon the system exceeds the capacity of the pumps. The tanks are equipped with automatic "altitude" valves that keep the tanks from overflowing. Systems are typically set up so that tanks contribute some water each day, which keeps fresh water circulating through the tanks.

Simple Loops

A simple loop will be defined by the following:

- There is exactly one inflow point and one outflow point.
- Exactly two paths exist between the inflow and outflow points.

Note that there is never more than one inflow or more than one outflow point (**Figure 3.38**). Notice that there are only two possible ways to get

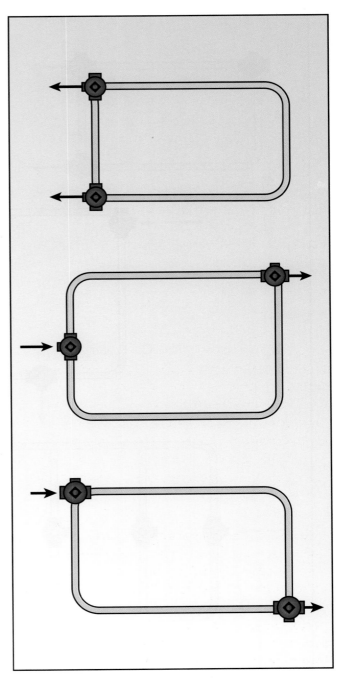

Figure 3.38 Examples of simple loops.

from the inflow point to the outflow point. Any deviation from the definition of a simple loop will give erroneous results when using the techniques presented in this chapter. For example, if there are three paths or two outflow points, the methods for calculating friction loss presented in this first section will not work.

A *complex loop* (which will also be referred to as a *grid* in this text) is a piping system characterized by one or more of the following:

- More than one inflow point exists.
- More than one outflow point exists.
- More than two paths exist between inflow and outflow points.

Notice that the existence of any one of the above is sufficient to characterize the system as complex. All three of the situations do not have to exist. Therefore, all of the systems illustrated in **Figure 3.39** qualify as complex.

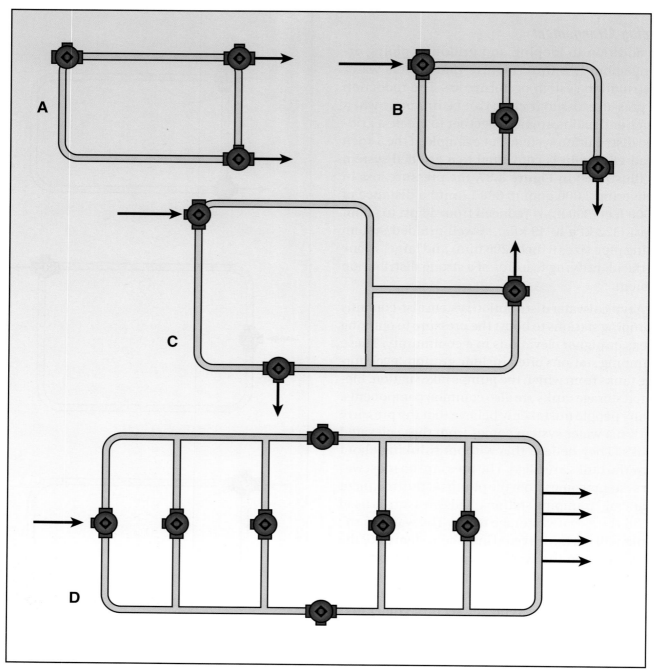

Figure 3.39 Examples of complex loops.

Valves

Valves are placed on water distribution systems to provide the ability to section and isolate portions of the system. This is a particularly important reliability feature. When a water main breaks, sufficient valves will necessitate only a small portion of the system being placed out of service while repairs are made. These are typically underground valves that require a special wrench, often called a "key," to operate the valve **(Figure 3.40)**. Ideally, these valves should be located on each branch at the intersection of mains (often referred to as a curb box) and no more than 500 feet (167 m) apart. There should also be a valve on every branch feeding a hydrant so that each hydrant can be isolated during repair or replacement.

Since these valves are underground, access is only possible through metal valve plates located at ground level **(Figure 3.41)**. Unfortunately, these access plates are often covered over with dirt or pavement, leaving valves inaccessible and their location hidden. A good maintenance program will require good water maps indicating valve location and conduct yearly valve operations to make sure the valves work properly and are completely open.

Valves for water systems are broadly divided into *indicating* and *nonindicating* types. An indicating valve visually shows whether the gate or valve seat is open, closed, or partially closed. Valves in private fire protection systems are usually of the indicating type. Two common indicator valves are the *post indicator valve* (PIV) and the *outside stem and yoke* (OS&Y) valve. The post indicator valve is a hollow metal post that is attached to the valve housing **(Figure 3.42, p. 112)**. The valve stem inside this post has the words *OPEN* and *SHUT* printed on it so that the position of the valve is shown. The OS&Y valve has a yoke on the outside with a threaded stem that controls the gate's opening or closing **(Figure 3.43, p. 112)**. The threaded portion of the stem is out of the yoke when the valve is open and inside the yoke when the valve is closed. These valves are most commonly used on sprinkler systems but may be found in some water distribution system applications.

Figure 3.41 A typical underground valve installation.

Figure 3.40 Valve keys are needed to close off underground valves.

Nonindicating valves in a water distribution system are normally buried or installed in manholes (**Figure 3.44**). These are the most common types of valves used on most public water distribution systems. If a buried valve is properly installed, the valve can be operated aboveground through a valve box. A special socket wrench on the end of a reach rod operates the valve.

Control valves in water distribution systems may be either gate valves or butterfly valves. Both valves can be of the indicating or nonindicating type. Gate valves may be of the rising stem or the nonrising stem type (**Figure 3.45**). The rising stem type is similar to the OS&Y valve. On the nonrising stem type, the gate either rises or lowers to control the water flow when the valve nut is turned by the valve key (wrench). Nonrising-stem gate valves should be marked with a number indicating the number of turns necessary to completely close the valve. If a valve resists turning after fewer than the indicated number of turns, it usually means that debris or other obstructions are in the valve. Butterfly valves are tight closing and they usually have a rubber or a rubber-composition seat that is bonded to the valve body (**Figure 3.46**). The valve disk rotates 90 degrees from the fully open to the tight-shut position. The nonindicating butterfly

Figure 3.42 The PIV shows the status of the valve through a small window.

Figure 3.43 The outside stem and yoke valve is open when the threaded portion of the stem is outside the yoke.

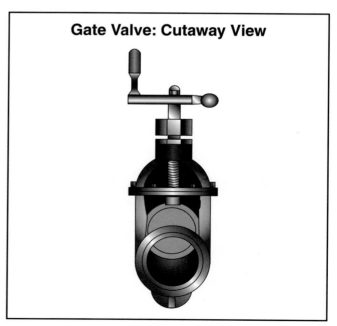

Figure 3.44 Nonindicating valves are not as easily accessed as other types of valves.

Gate Valve: Cutaway View

Figure 3.45 A gate valve opens and closes the waterway with a gate that moves up and down inside the valve.

type also requires a valve key. Its principle of operation provides satisfactory water control after long periods of inactivity.

The advantages of proper valve installation in a distribution system are readily apparent. If valves are installed according to established standards, it is usually necessary to close off only one or perhaps two fire hydrants from service while a single break is being repaired.

The advantage of proper valve installation is reduced if all valves are not properly maintained and kept fully open. A high level of friction loss will result if valves are only partially open. When valves are closed or partially closed, the condition may not be noticeable during ordinary domestic

Figure 3.46 The butterfly valve has a disk (baffle) that pivots within the waterway, interrupting the flow.

flows of water. As a result, the impairment will not be known until a fire occurs or until detailed inspections and fire flow tests are made. A fire department will experience difficulty in obtaining water in areas where there are closed or partially closed valves in the distribution system.

Hydrants

No matter how strong or reliable a water system is, it is of little value if the water is not accessible. It is through the fire hydrants that the fire department gains access to the water supply system. Fire hydrants are available in two basic types: dry barrel and wet barrel. The most common type of hydrant used in the United States is the dry-barrel type. As the name implies, there is no water inside the hydrant until the hydrant is turned on. This hydrant, therefore, has application in areas where freezing temperatures are expected. The valves in the hydrant keep water out of the barrel until it is opened, and drains located at the bottom of the hydrant allow the water to drain out when the hydrant is closed. When the hydrant is fully opened, the drain holes are closed off. If the hydrant is only partially opened, the drain holes will be partially open, permitting a pressurized stream of water to be discharged out the drains beneath the ground. This drainage will erode the area at the base of the hydrant. For this reason, dry-barrel hydrants should always be completely open when in use. Dry-barrel hydrants are easily recognized by the existence of the operating nut on the "bonnet" or top of the hydrant (**Figure 3.47, p. 114**).

The most common variety of dry-barrel hydrant has two 2½-inch outlets and one 4½-inch outlet (**Figure 3.48, p. 113**). The larger outlet is often called a "pumper" or "steamer" outlet because it is common for the fire department to attach its soft suction line to this outlet to feed the pumping apparatus. It is also common to see dry-barrel hydrants with only the two 2½-inch outlets and no pumper outlet. These outlets limit the capacity of the hydrant; therefore, hydrants of this type should not be used in industrial and commercial areas of a community. There are several other combinations of outlets that may be encountered, such as hydrants having only one or two pumper outlets and no smaller outlets. However, the first two described above are by far the most common.

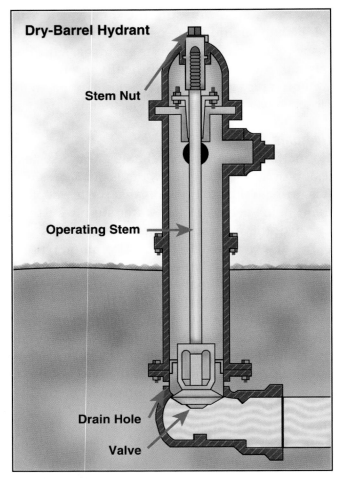

Figure 3.47 Dry-barrel hydrants utilize a long operating stem to keep the water well below ground when the hydrant is not in use.

Figure 3.48 A typical dry-barrel hydrant.

Wet-barrel hydrants may be found where freezing temperatures are not expected, such as in southern California, Arizona, or southern Florida. Water is up inside the hydrant at all times and would be subject to freezing in cold climates **(Figure 3.49)**. The operating nuts for turning the hydrant on extend through the side of the barrel, and there must be an operating nut for each outlet on the hydrant.

Whether they are wet barrel or dry barrel, hydrants are attached to the city main by a short piece of pipe called a "*branch*." This branch should be at least 6 inches (150 mm) in diameter. Branches 4 inches (100 mm) in diameter are common on older systems, but this size limits the capacity of the hydrants. Each branch should also have a valve so that the hydrant can be isolated for repair or replacement without shutting down the city main to which it is attached.

Proper hydrant distribution and location is an important feature of an accessible water supply system. A good rule of thumb is that there should be a hydrant at every intersection, and hydrants should not be spaced farther than 500 feet (166 m) apart. However, the International Fire Code bases hydrant spacing upon fire flow. It indicates that a 500-foot separation is appropriate only where fire flow requirements are 1,750 gpm (7 000 L/min) or less. A spacing as little as 200 feet (66 m) is indicated where the fire flow requirement is 7,500 gpm (30 000 L/min) or more. The pumper outlet should be positioned facing the street and the hydrant should be set high enough that the hydrant wrench can be turned a full revolution when removing the cap from the lowest outlet (usually the pumper outlet).

From a maintenance standpoint, grass and bushes should be kept trimmed away from hydrants so that they remain visible and accessible.

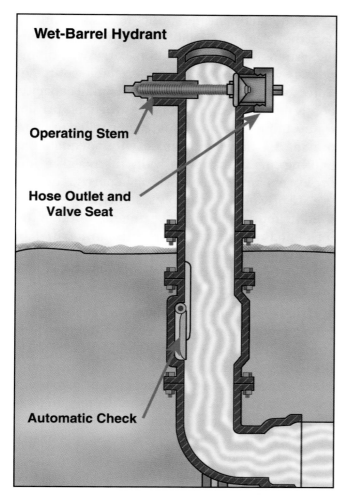

Wet-Barrel Hydrant

Operating Stem

Hose Outlet and
Valve Seat

Automatic Check

Figure 3.49 Wet-barrel hydrants have water right up to the discharge outlets when not in use.

Hydrant Class	Color	Flow
Class AA	Light Blue	1,500 gpm (6 000 L/min) or greater
Class A	Green	1,000-1,499 gpm (4 000 L/min-5 996 L/min)
Class B	Orange	500-999 gpm (2 000 L/min-7 996 L/min)
Class C	Red	less than 500 gpm (2 000 L/min)

Table 3.4
Hydrant Color Codes

Reprinted with permission from NFPA 291, *Recommended Practice for Fire Flow Testing and Marking of Hydrants.* Copyright© 1995 National Fire Protection Association, Quincy MA 02269. This reprinted material is not the complete and official position of the National Fire Protection Association on the referenced subject, which is represented only by the standard in its entirety.

Hydrants should be flushed every year to make sure that they work and to clean sedimentation from the pipes. The Insurances Services Office (ISO) requires flushing twice a year for full credit. It is also a good idea to measure the flow at each hydrant every year. By comparing the flows at each hydrant year by year, any deterioration of or obstruction to the water supply system can be easily detected.

In many communities, fire hydrants are color coded to reflect the gpm capacity of the fire hydrants. The system recommended in NFPA 291, *Recommended Practice for Fire Flow Testing and Marking of Hydrants* is shown in **Table 3.4**. Each of the flows shown in the table would be available from the indicated hydrant at a residual pressure of 20 psi (140 kPa).

Private Water Supply Systems

In addition to the public water supply systems that service most communities, fire department personnel must also be familiar with the basic principles of any private water supply systems that are within their response jurisdiction. Private water supply systems are most commonly found on large commercial, industrial, or institutional properties. They may service one large building or a series of buildings on the complex. In general, the private water supply system exists for one of the three following purposes:

- To provide water strictly for fire protection purposes
- To provide water for sanitary and fire protection purposes
- To provide water for fire protection and manufacturing processes

The design of private water supply systems is typically similar to that of the municipal systems described earlier in this chapter. Most commonly, private water supply systems receive their water from a municipal water supply system. In some cases, the private system may have its own water supply source independent of the municipal water distribution system **(Figure 3.50, p. 116)**.

Sometimes a property may be served by two sources of water supply for fire protection: one from the municipal system and the other from a private source. In many cases, the private source of water for fire protection provides nonpotable (not for

Figure 3.50 Although many private water systems receive their water from a municipal water system, some have their own source of water.

drinking) water. When this is the case, adequate measures must be taken to prevent contamination caused by the backflow of nonpotable water into the municipal water supply system. There are a variety of backflow prevention measures that can be employed to avoid this problem. Some jurisdictions do not allow the interconnection of potable and nonpotable water supply systems. This means that the protected property is required to maintain two completely separate systems.

Almost all private water supply systems maintain separate piping for fire protection and domestic/industrial services. This is in distinct contrast to most municipal water supply systems in which fire hydrants are connected to the same mains that supply water for domestic/industrial use. Separate systems are cost prohibitive for most municipal applications but are economically practical in many private applications. There are a number of advantages to having separate piping arrangements in a private water supply system, including the following:

- The property owner has control over the water supply source.

- Neither system (fire protection or domestic/industrial) is affected by service interruptions to the other system.

Keep in mind that private water supply systems that rely solely on the municipal water distribution system as their water supply source are subject to service interruptions if the municipal system experiences a failure.

Fire department personnel must be familiar with the design and reliability of private water supply systems in their jurisdiction. Large, well-maintained systems may provide a reliable source of water for fire protection purposes. Small capacity, poorly maintained, or otherwise unreliable private water supply systems should not be relied upon to provide all the water necessary for adequate fire fighting operations. Historically, many significant fire losses can be traced, at least in part, to the failure of a private water supply system that was being used by municipal fire departments working the incident. Problems such as the discontinuation of electrical service to a property whose fire protection system is supplied by electrically driven fire pumps have resulted in disastrous losses.

If there is any question about the reliability of a private water supply system or of its ability to provide an adequate amount of water for a large-scale fire fighting operation, the fire department should make arrangements to augment the private water supply. This may be accomplished by relaying water from the municipal water supply system or by drafting from a reliable static water supply source close to the scene.

Elevated tanks are used to stabilize or balance the pressures on a water system at times of peak demand. As long as the pumps supplying a system can keep up with the demand, the water in the tanks will not be used. However, once the demand upon a system becomes so great that the pumps cannot keep up, the system pressure will begin to drop. When the pressure drops to a point where it cannot keep the tank full, the tank will begin to add water to the system.

One of the greatest peak demand periods is when a major fire occurs in a community. For this reason, the existence of elevated storage becomes an important fire protection feature. This chapter addresses the analysis of combining an existing water supply with elevated storage to determine the total supply available.

In times past it was very common to see industrial plants with a private elevated tank on site as a backup for the public water supply. However, there are several problems associated with providing fire protection water supplies from elevated tanks. The most important of those is the limitation of pressure available from an elevated tank. A tank has to be 100 feet (33 m) high to generate a pressure of only 43 psi (301 kPa). This pressure is not adequate to meet many modern fire system requirements. For example, Early Suppression Fast Response sprinklers used for warehouse protection require at least 50 psi (350 kPa) at the most hydraulically demanding sprinkler. System demands for other high-challenge occupancies would also require tanks to be of impractical heights to provide adequate pressure. For this reason, it makes more sense to use ground-level storage and a fire pump for the water supply redundancy required at many highly protected facilities.

Elevated tanks provide a very reliable water supply. It is only necessary to keep the tanks full and the valves open to be assured that water will flow when needed. However, tanks do require considerable maintenance and often require protection against freezing. They are also an attractive nuisance and are frequent targets of vandalism and malicious mischief.

Water Supply Requirements for Standpipe and Hose Systems

For standpipe and hose systems, the water supply requirements are more straightforward. NFPA 14, *Standard for the Installation of Standpipe and Hose Systems,* provides specific requirements. NFPA 14 defines three classes of standpipe service with different requirements for each class:

- **Class I service** essentially consists of a system intended for use by the fire department. Class I systems are required to have 500 gpm (2 000 L/min) available for the first standpipe plus 250 gpm (1 000 L/min) for each additional standpipe. A 1,250 gpm (5 000 L/min) maximum is specified. A residual pressure of 100 psi (700 kPa) is required at the most hydraulically demanding hose connection. The most hydraulically demanding hose connection would be the one with the greatest amount of pressure lost to friction and elevation when delivering 500 gpm (2 000 L/min).

- **Class II service** is intended to be used primarily by building occupants as first aid fire appliances. This class requires only 100 gpm (400 L/min),

with no increase for multiple standpipes. The pressure requirement is 65 psi (455 kPa) at the most remote connection with 100 gpm (400 L/min) flowing. Since the Class II service is for building occupants, the pressure and flow must be available independent of fire department pumping apparatus. This means that for most buildings with the Class II service a pump will be needed.

- **Class III service** is a combination of Classes I and II. Class III water supply requirements are identical to Class I requirements. NFPA 14 requires automatic water supplies. The support of the systems through fire department connections is considered an auxiliary supply. Even though this requirement has debatable merits for the Class I service, alternate design approaches should be undertaken only after obtaining permission from governing authorities.

Water supplies for standpipe systems also provide water for sprinkler systems. In these combined systems, if the buildings are fully sprinklered the sprinkler demand does not have to be added to the standpipe demand. The standpipe demand alone is used. However, if the building is only partially sprinklered, the sum of the standpipe demand and the sprinkler demand must be used.

This section was not intended to address the design of standpipe and hose systems, only the water supply requirements. The detailed design specifications may be found in NFPA 14. However, it should be noted that 1½-inch (38 mm) hose stations are rarely installed in modern buildings and those that exist are rapidly being removed. Fire departments will not use these small hoselines because building owners historically have not maintained the hose. To use poorly maintained hose could jeopardize the safety of firefighters. In addition, the presence of hose stations is evidence that building owners expect building occupants to use the hose. Under federal law (OSHA) this means building occupants must be trained in the use of small hoselines. To avoid the complications and expense of hose maintenance and employee training, most hoselines have been removed.

It should also be noted that the small hoseline systems required by the NFPA warehousing standards (231, 231C, etc.) are not intended to be systems complying with NFPA 14. Because these are separate standards, pressure and flow requirements for warehouse hose stations are entirely different from those dictated by NFPA 14.

NOTE: For more information about standpipe systems, see Chapter 5 of this manual.

Water Supply Requirements for Automatic Sprinkler Systems

There are two recognized design techniques for automatic sprinkler systems. The older, more traditional approach is the pipe schedule design technique. The more modern method involves hydraulic calculations. The pipe schedule method will be discussed first.

NOTE: Sprinkler systems are discussed in more detail in Chapter 6 of this manual.

Pipe Schedule Systems

Pipe schedule systems are those in which the pipe sizing is based upon the number of sprinklers supplied, as dictated by tables in NFPA 13, *Standard for the Installation of Sprinkler Systems*, Chapter 6. Systems designed in this manner have been around for more than a century, and their performance record has been outstanding. From a fire control standpoint, there is nothing wrong with pipe schedule design. However, certain severe hazards require more water than the typical pipe schedule system can deliver. From an economical standpoint, a hydraulic analysis might reveal that smaller pipe than required by the pipe schedules can provide adequate protection. Under proper occupancy considerations, pipe schedule design provides excellent protection, but perhaps not the most economical system.

Hydraulically Designed Water-Based Extinguishing Systems

For the hydraulically designed automatic sprinkler systems, mathematical calculations are used to arrive at specific flow and pressure requirements. The water supply demands for other water-based systems such as foam systems, water spray systems, or foam-water sprinkler systems will be determined through minor variations in the hydraulic calculation technique for sprinkler systems. For example, with the deluge water spray systems, the primary variation is that all nozzles are open and will deliver water. With the conventional sprinkler system design, only a few sprinklers are considered to be open.

With the foam system, differences include the use of the Darcy-Weisbach technique for calculating friction loss before the foam concentrate is mixed with water. The number of gpm required is also based on square feet of surface to be covered. Other minor differences exist. However, design or evaluation of these systems requires knowledge of hydraulic calculations.

Duration of Water Supplies

The required duration of fire protection water supplies is essential information when designing water storage facilities. The required duration of sprinkler system water supplies is given in **Table 3.5** or in specific design standards such as NFPA 231, *Standard for General Storage*.

Where a duration range is given, the higher number is typically recommended unless system alarms transmit directly to the fire department. For example, if a fire pump taking water from a tank supplies water for sprinklers in a light hazard building where the demand is considered to be 500 gpm (166 L/min), the required duration of the water

Occupancy Classification	Minimum Residual Pressure Required (psi)	Acceptable Flow at Base of Riser (Including Hose Stream Allowance) (gpm)	Duration (minutes)
Light Hazard	15	500 – 750	30 – 60
Ordinary Hazard	20	850 – 1,500	60 – 90

Table 3.5
Water Supply Requirements for Pipe Schedule Sprinkler systems

Note: For SI units, 1 gpm – 3.785 L/min; 1 psi – 0.0689 bar.

Reprinted with permission from NFPA 13-2002, Installation of Sprinkler Systems, Copyright© 2002 National Fire Protection Association, Quincy, MA 02269. This reprinted material is not the complete and official position of the National Fire Protection Association on the referenced subject, which is represented only by the standard in its entirety.

supply would be 60 minutes with a local alarm only. The tank volume would have to be at least 30,000 gallons (60 min x 500 gpm).

The duration of water supplies for exterior manual fire fighting is recommended to be 2 hours for up to 2,500 gpm (10 000 L/min) and 3 hours for larger flows. The required duration for standpipe and hose systems is 30 minutes for all Class I, II, and III service.

NOTE: For more detailed information on water supply analysis, see the Fire Protection Publications *Fire Protection Hydraulics and Water Supply Analysis* manual.

Summary

Water continues to be the most readily available and plentiful fire extinguishing agent. Knowing that water is available is not enough, however; it is also important to know how water extinguishes fire in the first place in addition to advantages and disadvantages of using water. In any kind of distribution system, a municipal or private water supply system or standpipe or sprinkler system — water must be available at sufficient quantities and adequate pressures to be effective. Personnel who must inspect or use these systems need to know how water is distributed as well as any physical and design factors that can positively or negatively affect its availability. The work that is done to keep these systems operating effectively is critical because their purpose is to save lives and property.

Fire Pumps

1.   Describe the three most common types of fire pumps.

2.   List the types of pump drivers that are acceptable for use with fire pumps, including their advantages and disadvantages.

3.   Describe the operation of a pump controller.

4.   Recognize and identify all pump components and accessories required for the installation of a fire pump.

5.   Describe the types of pipes and fittings that are required for the installation of fire pumps.

6.   Describe the function of a relief valve on a fire pump.

7.   Describe the component arrangement for the installation of a fire pump.

8.   Describe the testing procedures to be used when testing a fire pump.

9.   Describe routine maintenance procedures for fire pumps.

Chapter 4
Fire Pumps

A fire pump is a fixed pump that supplies water to a fire suppression system. The main function of a fire pump is to increase the pressure of the water that flows through it. Usually a fire pump is needed to supply a sprinkler or standpipe system because the available water supply source, such as an elevated tank or ground storage tank, does not have adequate pressure to meet the demands of the fire suppression system. Water is available to a fire pump from sources such as municipal water mains, wells, storage tanks, and reservoirs.

The most common type of water supply for fire protection comes from a public water supply system. These water supplies can be adequate in every respect and reasonably reliable. However, it is common to find industrial or commercial facilities in locations where there is no public water supply or where the supply is too weak to supply fire protection needs. Two such examples are as follows:

- The public water supply can deliver enough gallons per minute or liters per minute, but the pressure is too low to meet the fire protection demand.

- The public water supply may or may not exist and can provide neither an adequate number of gpm (L/min) nor adequate water pressure.

The earliest fire pumps were positive-displacement types and used either rotary gears or pistons; modern fire pumps are centrifugal pumps. The centrifugal force of an impeller or the rotating vanes of a turbine add pressure to the water. This chapter is limited to the discussion of the most common types of pumps: horizontal split-case pumps, vertically mounted split-case pumps, and vertical-shaft turbine pumps **(Figure 4.1)**.

Figure 4.1 A horizontal shaft fire pump.

This chapter addresses the types of fire pumps commonly installed in business and industrial facilities to support existing water supplies. A discussion of the valves, piping, components, and accessories that make up a standard pump installation is also included. Criteria for the installation, testing, and maintenance are listed in NFPA 20, *Standard for the Installation of Stationary Pumps for Fire Protection,* 2003 Edition.

NOTE: Fire pumps are tested and listed by Underwriters Laboratories Inc. and may be tested and approved/listed by FM Global. The companies manufacturing fire pumps and descriptions of their equipment are contained in lists published by these two organizations.

Common Fire Pump Types
Horizontal Split-Case Pumps

In split-case pumps, the casing in which the shaft and impeller rotates is split in the middle and can be separated, thus exposing the shaft, bearings, and impeller. This provides easy access for repair or replacement of internal components (**Figure 4.2**). There are two types of split-case pumps: horizontally mounted and vertically mounted. The horizontal split-case centrifugal fire pump is the most common of the two types and is acceptable for use where water can be supplied to the pump under some pressure. This is because the pump is not self-priming. The pump must be supplied by a public water supply system or a tank located above the level of the pump-intake port. Otherwise, the pump would need to be primed, which is unacceptable where automatic pump operation is required.

In the horizontal split-case pump, the water pressure is increased due to the operation of a rotor inside the pump casing. This rotor is called an *impeller* (**Figure 4.3**). Water is fed into the center, or eye, of the impeller from one or both sides and then thrown to the outer edges by the rotation of the impeller (**Figure 4.4**). As the water is forced to the outer edge of the impeller, the energy or pressure increase depends upon how fast the impeller is turning and the diameter and design of the impeller. The impeller turns on a shaft that is usually driven by an electric motor or a diesel engine. Though relatively rare, pumps driven by a steam turbine may be encountered.

Fire pumps are referred to as single-stage or multiple-stage. A single-stage pump is one with a single impeller. Most horizontal split-case pumps

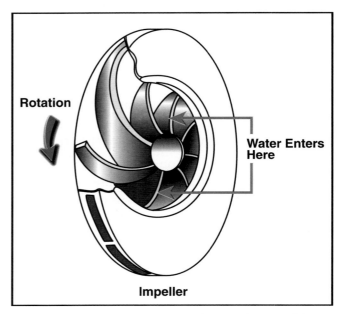

Figure 4.3 Water enters the pump through the eye of the impeller.

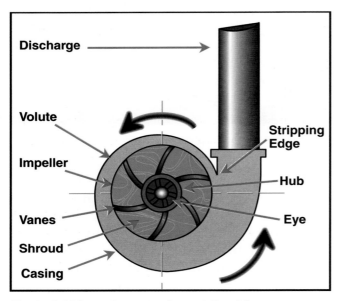

Figure 4.2 The major parts of a centrifugal fire pump.

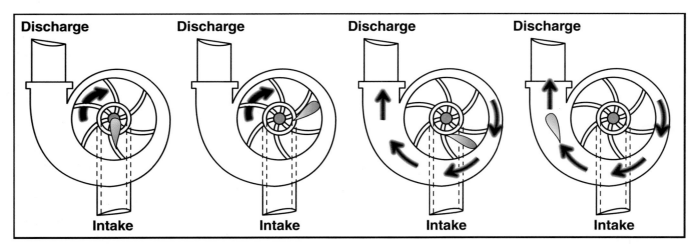

Figure 4.4 These schematics trace the path of water through the centrifugal fire pump.

are single-stage pumps. Multiple-stage pumps have the ability to deliver higher pressures and are commonly used for standpipe water supplies in high-rise buildings.

Fire pumps that are as small as 25 gpm (8.3 L/min) are listed by the Underwriters Laboratories standard; however, it is rare to encounter pumps rated at less than 500 gpm (2 000 L/min). From a practical standpoint, the only time a pump smaller than 500 gpm (2 000 L/min) would be adequate for fire protection purposes is with small, light-hazard sprinkler systems and with a Class II standpipe and hose system, which is composed of hose cabinets and 1½-inch (38 mm) hoselines. The most common pumps sizes are 500, 750, 1,000, and 1,500 gpm, (2 000 L/m, 3 000 L/min, 4 000 L/min, and 6 000 L/min), but pumps as large as 4,500 gpm (1 800 L/min) are available.

There is no standard pressure rating, but single-stage horizontal split-case centrifugal pumps may be purchased with pressure ratings as low as 40 psi (280 kPa) and as high as 290 psi (2 030 kPa). No standard fire pump is permitted to have a pressure rating less than 40 psi (280 kPa).

As its name implies, the horizontal split-case pump has its shaft oriented horizontally. A grease fitting is provided for the bearing at each end of the shaft to allow proper lubrication, which must be performed at regular intervals. There are also fiber packings on each end of the shaft, and these are inserted in the packing gland, which is a kind of sleeve (**Figure 4.5**). Their function is to seal the shaft and prevent excess water leakage around the shaft. They are intended to be water cooled and lubricated, so a small leakage is required; about one drop per second is generally considered adequate. Too much or too little leakage requires tightening or loosening of the packing gland.

Additional important components for pump maintenance are case wearing rings. These are located at the interface between the pump casing and the impeller on each side of the impeller. Their function is to prevent wearing of the pump casing. Since these wearing rings suffer all the wear, they must be checked regularly and replaced periodically.

Every fire pump must have a driver to turn the pump impeller and an electric controller to turn the pump on and off. The fire pump must be installed in a system of piping and valves. More attention is given to these components in the following sections.

Vertically Mounted Split-Case Pumps

Even though the name implies that a horizontal split-case pump will be installed with the drive shaft in the horizontal position, there is a modern variation. In some installations where the pump is driven by an electric motor, the motor may be installed on top of the pump. This is a standard installation technique and saves floor space. This type of pump is referred to as the vertically mounted split-case pump (**Figure 4.6, p. 126**).

The vertically mounted pump is basically the same as the horizontal split-case pump and has identical applications. The impeller rotates on a shaft within the pump casing. The water is discharged in a direction perpendicular to the shaft, and the shaft rotates on ball-type bearings at each end.

Vertical-Shaft Turbine Pumps

The vertical-shaft turbine centrifugal fire pump was originally designed to pump water from wells. It still has application where the water supply is from a nonpressurized source, including wells, ponds, rivers, and underground storage tanks. Vertical-shaft pumps never require priming because the rotating turbines are positioned down inside the water source. Vertical-shaft pumps ordinarily

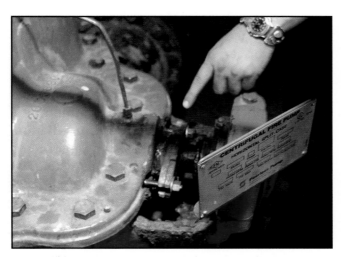

Figure 4.5 Packing glands are designed to prevent excess water leakage around the shaft.

Figure 4.6 The vertically mounted split-case pump saves floor space.

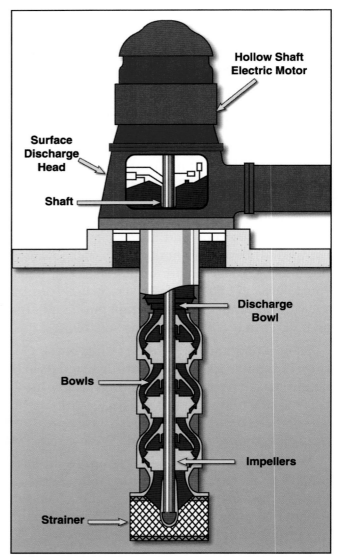

Figure 4.7 Shown are the principal parts of a vertical turbine fire pump assembly.

have more than one impeller and are therefore multistage pumps. The number of impellers is determined by the desired pressure rating; the more impellers there are, the greater the pressure that will be developed. As the water exits one impeller, it enters the next, and so on until it is discharged into the fire suppression system piping.

Figure 4.7 illustrates the essentials of the vertical-shaft pump. Note that each turbine sits in a slight enlargement of the casing (referred to as a bowl). Each bowl has a wear ring to prevent wearing of the bowl itself. The shaft can be either water lubricated or oil lubricated. A strainer is located at the bottom to keep fish, snails, leaves, and other objects out of the pump.

The shaft is turned by an electric motor, a steam turbine, or a diesel engine. There is also fiber packing at the top of the shaft to seal the shaft against water leakage. The fiber packing is

contained inside a packing gland, which can be either tightened or loosened to obtain the proper packing lubrication.

In recent years, an application of the vertical-shaft turbine pump to support pressurized water supplies has become more common. It is commonly referred to as a can-type installation. Essentially, a cylinder or canister is constructed and the city water supply feeds this canister. The vertical-shaft pump is simply set into this canister and is then able to boost the pressure of the incoming water supply (**Figure 4.8**).

The reason that vertical-shaft pumps are available with higher pressure ratings than horizontal-shaft pumps is because some of the pressure

Figure 4.8 The vertical-shaft pump is designed to boost the pressure of the incoming water supply.

developed by the vertical-shaft pump must be expended in simply getting the water up to ground level. How much pressure is lost in bringing the water to the surface depends upon how far below grade the impellers are located.

Even though the impellers of a vertical-shaft pump are located below ground level inside the water source, the driver and control panel will be accessible above ground. Vertical-shaft pumps are usually driven by an electric hollow-shaft motor mounted above the pump or by a diesel engine through a right-angle gear drive.

Both horizontal and vertical-shaft pumps may have a pressure rating in terms of feet of head rather than in pounds per square inch (psi). Therefore, a pump that is rated at 231 feet of head is equivalent to a pump that is rated at 100 psi (700 kPa), since 231 feet x 0.433 psi/ft is equal to 100 psi. Vertical turbine pumps are available with discharge pressure ratings of up to 500 psi (3 500 kPa). Some pumps are designed to pump seawater. If these pumps are rated in feet of head, 0.445 psi/ft should be used to convert to psi.

Pump Drivers

The pump is just one component of a fire pump installation. In the next two sections, two of the most important components – drivers and controllers – are discussed. The drivers are the engines or motors used to turn the pump. The controllers are the electrical control panels used to switch the pumps on and off and to control their operation.

Currently, there are three types of drivers that are acceptable for use with fire pumps: electric motors, diesel engines, and steam turbines. Before

1974, other kinds of internal combustion engines, such as gasoline, natural gas, or propane engines were permitted. These engines used a spark plug ignition system and were found to be less reliable in starting than diesel engines. Some of these types of engines can still be found on older installations, but modern pump installations should have diesel engines if an internal combustion driver is used.

The pump driver must have enough power to turn the pump at rated speed under all required load conditions, which include pumping at churn and pumping at 150 percent of rated capacity. A pump is said to be operating at *churn* or *shut-off* when it is running but all discharges are closed. The horsepower ratings of the pump driver are commonly found on an information plate on the engine. The horsepower varies according to the speed at which the driver is operating. For this reason, it is very important to keep a pump driver well maintained and properly adjusted. Otherwise, the pump cannot be expected to meet its performance specifications.

Electric Motors

Electric motors have long been a dependable source of power for driving centrifugal fire pumps. Electric motors used on fire pumps were not originally designed for that purpose; however, all electric motors must meet the requirements of the National Electrical Manufacturers Association (NEMA). The motor must have adequate horsepower to drive the fire pump. The required pump horsepower is determined by the pump capacity (gpm), the net pressure (discharge pressure minus the incoming pressure), and the pump efficiency. For a 1,000

Figure 4.9 Placing the electric motor beside the pump helps ensure proper shaft alignment.

gpm (4 000 L/min) pump rated at 100 psi (700 kPa), a motor of about 80 hp would be needed. Electric motors powerful enough to power fire pumps use a great deal of electricity and may require a larger electrical service to the building than would be needed otherwise. For the horizontal split-case pump, the electric motor is most commonly located on a framework beside the pump to ensure proper shaft alignment **(Figure 4.9)**. Split-case pumps are also available for vertical installation with the motor located on top to save floor space.

The electric motors driving vertical-shaft pumps are mounted vertically on top of the pump turbine shaft. They must be hollow-shaft motors with an anti-reverse ratchet that prevents the motor from turning backwards **(Figure 4.10)**. If not equipped with this feature, the motor can turn in the wrong direction when water drains down the column after the pump is turned off or if the motor is wired incorrectly. The motor must also be equipped with thrust bearings to carry the weight of the pump plus withstand the thrust of the water.

One of the most important considerations is the size of the motor relative to the pump. The motor must have adequate power to turn the pump at its

Figure 4.10 The impeller assembly of this vertical-shaft pump draws water from the storage tank below. *Courtesy of Southwest Loss Control, Inc.*

rated speed so that the pump can deliver its rated flow and pressure. A motor of ample size must be provided. The adequacy of the electric motor will become apparent during acceptance tests when the motor is required to turn the pump under various load conditions.

A decided advantage of the electric motor is the relatively small amount of maintenance required. Proper lubrication of the motor bearings in accordance with manufacturer's instructions is all that is required. A disadvantage of the electric motor involves reliability. Storms, fires, or other accidents involving power lines, transformers, or substations can leave motors without power and fire pumps useless. Wiring installations providing power for the electric motors and the controllers are required to comply with the provisions of NFPA 70, *National Electrical Code®*.

Electrical Power

It is beyond the scope of this text to discuss the details of the electrical power system; however, a few points are worth noting. The pump motor should be wired so that power to the plant or facility can be shut off without eliminating the power to the pump motor. This can be accomplished by providing separate power services or by locating the fire pump service connection ahead of all plant service equipment. In addition, power lines supplying electricity to the pump should not run through the protected building unless they are inside fire-resistive enclosures.

Voltages in excess of 600 (AC) should not be used for fire pump service. Each electric motor should have an information plate giving the current and voltage ratings, horsepower, rpm, and service factor. This data is useful in determining the acceptability of pump performance during the tests, which are described later in this chapter.

Electric motor driven pumps are often neglected. This is partly due to their simplicity and to the fact that very little maintenance is required. However, these pumps should be turned on weekly and permitted to run for at least 10 minutes.

Diesel Engine Drivers

The diesel engine is a common and reliable method of powering fire pumps. Although it is usually more expensive than either gasoline or electric drivers, the diesel driver may be a better choice because it does not rely on external power. Electrically driven pumps are simpler and require less maintenance than diesel pumps **(Figure 4.11)**. The diesel engine, however, has proven to be the most dependable of all of the internal combustion engines and is currently the only kind of internal combustion engine considered acceptable for fire protection applications.

Diesel engines are listed by Underwriters Laboratories Inc. and approved by FM Global for fire protection applications. This means that not all diesel engines are acceptable for driving fire pumps. If a diesel engine driver is used on a fire pump application, the testing agencies require that the engine be equipped with overspeed shutdown devices, tachometers, oil pressure gauges, and temperature gauges. Look for the UL or FM label; it is the manufacturer's responsibility to properly match the driver to the pump.

Figure 4.11 A typical diesel motor driven fixed fire pump installation. *Courtesy of ConocoPhillips.*

Engine Power

If the engine is operated at speeds that are too high, the pump can develop excessive water pressure that damages both the engine and the pump. If the engine operates too slowly, the pump will not develop rated pressures. The engine is therefore required to have an adjustable governor to maintain engine speed within a 10 percent range. The governor is required to be set to maintain rated pump speed at maximum pump load. The engine is also required to have an over-speed shutdown device. If the governor fails to limit pump speed, this device will shut down the engine to a speed approximately 20 percent above the rated engine speed. The shutdown device is required to be manually reset and supervised to send a trouble signal to the control panel until it is reset.

Gear Drives

The diesel engine is connected to a horizontal split-case pump by a flexible coupling. In the horizontal split-case pump, there is usually no gearing between the engine and the pump, and the pump speed will be identical to the engine speed. The vertical-shaft pump is driven by a diesel engine through a right-angle gear drive with universal joints. It is a geared connection and the primary purpose is to change the drive angle; however, it is common to find a gear ratio that will provide a pump speed different from the operating speed of the engine. Proper evaluation of pump performance requires a careful check for a gear ratio. The gear drives must also be equipped with the anti-reverse ratchet feature and the thrust bearings previously described for the electric motor driven vertical-shaft pumps.

Engine Requirements

The following instruments should be placed on a panel securely fastened to the engine at an accessible location.

- Tachometer – to indicate revolutions per minute (rpm) of the engine
- Oil pressure gauge – to indicate the pressure of the lubricating oil
- Water temperature gauge – to indicate the temperature of the water in the engine jacket

A speed-sensitive switch signals when the engine is running. Engines may be started electrically by storage batteries, compressed air, or hydraulics, but batteries are the most common method for starting the engine. For reliability, each battery unit must have sufficient capacity to maintain an engine cranking speed through six consecutive cranking cycles of 15 seconds cranking and 15 seconds resting.

Two means of recharging the batteries must be provided. One means of recharging is to use the generator or alternator that comes with the engine; another is to use an automatic charger that takes power from an alternating current source. The battery charger is incorporated into the design of the controller and must be capable of fully recharging the batteries within 24 hours. The batteries should be located so that they are not subject to flooding, mechanical damage, extreme temperature variations, or vibration. They should be readily accessible for easy servicing.

If a diesel engine is located in an environment that is subject to flammable vapors, starting the engine through the ordinary electric starting motor could create a fire or explosion hazard. For this reason, pneumatic and hydraulic starting is available. This is accomplished by forcing a compressed gas or water through a turbine that turns the engine.

For personnel safety, diesel engine exhaust fumes should be piped to the outside of the building. Inspect the integrity of the exhaust system regularly for leaks inside the pump room. The exhaust pipe should not be located close to combustible materials.

Cooling System

Diesel engines are water cooled and make use of closed-circuit-type cooling systems. The basic components of the system include a water pump driven by the engine, a heat exchanger, and a reliable device for regulating the water temperature in the engine jacket. The heat exchanger works by taking water from the discharge side of the pump to cool the water in the engine. The exchange of heat from the engine water to the pump discharge water takes place in the heat exchanger. The provision of water from the discharge side of the pump requires a special piping arrangement including bypass line, valves, pressure regulator, and strainers.

Fuel Storage

Diesel fuel is not as volatile as gasoline, but it can still be dangerous and must be handled carefully. Safe storage, transmission, and adequate quantities must be provided. Any exposed fuel lines must be protected against mechanical damage. Fuel piping should be rigid except where the fuel line connects to the engine. At that point, flame-resistant flexible hose is required.

For environmental protection, containment should be provided to prevent runoff from any leakage from the tank. This is commonly accomplished through construction of a small containment dike around the base of the tank. Codes and ordinances usually require the containment volume to be at least as large as the total volume of the tank. Other containment methods include sloping the floor to direct leaks to a containment basin or placing a sill around the pump room.

Maintenance

Diesel engines require more care and maintenance than electric motors. To ensure reliability, start and operate the engine for 30 minutes every week to ensure that it runs smoothly at the rated speed. Keep the engine clean, dry, and well lubricated. Follow the manufacturer's instructions in scheduling oil changes, but the interval between changes should not exceed one year.

Examine and test the batteries regularly to make sure they are fully charged. With the automatic battery charger, it is easy to assume that the batteries will always be in good shape. If more water is needed in the batteries, only distilled water should be used and the plates should be kept submerged at all times. Fuel tanks should be kept free of water and foreign materials. It would be helpful for building management to provide a maintenance chart or checklist to ensure that thorough inspections are made.

As the temperature drops, diesel engines become increasingly difficult to start. For this reason, the temperature of the pump room should be maintained at 70°F (20°C). If this is not practical, automatic heaters can be used to keep the engine warm. If these maintenance procedures are consistently followed, the diesel-driven fire pump can be relied upon to provide water for fire protection systems.

Steam Turbines

Although not common, some fire pumps are driven by steam. NFPA 20 lists steam turbines as an acceptable type of pump driver. Some of the older installations used piston-type reciprocating steam engines. Both horizontal- and vertical-shaft pumps can be driven by the steam turbine. The only feasible application of the steam-driven pump is when an uninterruptible supply of steam is available in sufficient quantities and at sufficient pressure. Otherwise, economic considerations would dictate use of the electric- or diesel-driven equipment.

Pump Controllers

The controller governs the operation of the pump, namely, it turns the pump on and off. The controller can be designed to operate the pump automatically by use of microprocessors, simple electronic circuits, or manual operation. The controller is usually in a large red cabinet that is against or hanging on the wall. Because it contains high-voltage wiring, only experienced maintenance personnel should service it.

WARNING!
Only experienced maintenance personnel should service the pump controller because it contains high-voltage wiring.

Controllers for Electric Motor-Driven Pumps

The controller should be tested and listed for fire protection use by one of the nationally recognized testing laboratories. The controller should be located inside the pump room and as close to the pump as possible. It should be protected from water discharge, and all current-carrying parts of the controller should be at least 12 inches (300 mm) off the floor **(Figure 4.12, p. 132)**. The main parts of the controller are discussed in the sections that follow.

Number 1 in Figure 4.12 is the *circuit breaker*. The function of the circuit breaker is to provide overcurrent protection by opening if too many amps are being drawn. The circuit breaker should be accessible and operable from the outside of the controller. Proper load rating of the circuit breaker is important. It should allow at least 115 percent of the rated full load current without

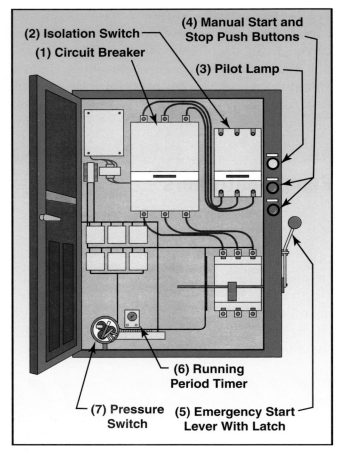

(2) Isolation Switch
(1) Circuit Breaker
(4) Manual Start and Stop Push Buttons
(3) Pilot Lamp
(6) Running Period Timer
(7) Pressure Switch
(5) Emergency Start Lever With Latch

Figure 4.12 All electric fire pump controllers have the same basic features.

tripping and also permit normal starting of the motor without tripping.

Number 2 is the *isolation switch,* which is located between the power supply and the circuit breaker. In some controllers, the isolation switch and the circuit breaker are interlocked so that the isolation switch cannot be operated with the circuit breaker closed. In other cases, the operating handle of the isolation switch is equipped with a spring latch that requires the use of both hands to operate. The isolation switch lever is always located outside the control panel.

CAUTION!
Because of the potential for damage to electrical equipment, the circuit breaker should always be opened before the isolation switch is operated.

Number 3 is a *pilot lamp.* It may be in various locations, but every controller should have one. It should be visible without getting inside the controller. Its purpose is to indicate when power is available to the pump control panel.

Number 4 is the manual *start and stop buttons.* They are standard controller components and located on the outside of the controller enclosure. Manual starting and stopping of the pump can be accomplished by using these buttons.

Number 5 is an *emergency start lever with latch.* This switch can be latched in the operating position and provide for continuous nonautomatic operation. This operation is independent of the timer or automatic starting mechanisms.

Number 6 is the *running period timer.* The timer is used to shut off the motor after starting causes have returned to normal. The running period timer should be set for at least 10 minutes. Therefore, if a pump is automatically controlled, it will operate for at least 10 minutes after being turned on unless manually shut down.

Number 7 is the *pressure switch.* The most common method of automatically turning on a fire pump is by using a pressure switch connected to the fire protection water supply system on the discharge side of the pump. The pressure switch is set to close the circuitry and turns on the fire pump if water pressure on the system drops below the switch setting. Such a pressure drop is usually caused by a sprinkler opening or by hoselines being operated somewhere in the system.

The pressure switch is adjustable, with independent high- and low-pressure settings. When the pressure on the system drops to the low-pressure setting, the pump is turned on. When the pressure is restored to normal, the high-pressure setting turns the pump off. The hydrostatic pressure of a sprinkler or standpipe system must be taken into account when setting the low-pressure setting. The low-pressure setting must be greater than the hydrostatic pressure of the system or the pump will not start. Ordinarily, the high-pressure setting should be about 10 pounds below shut-off or churn pressure and the low-pressure setting about 20 psi (140 kPa) below the rated pump pressure.

Consistent pressure fluctuations of sufficient magnitude on the fire protection system may require the use of a pressure maintenance pump (sometimes called a jockey pump) to prevent false pump starts. It should be noted, however, that if the pump is the sole source of water supply to the fire protection equipment, the pump should be set up for automatic start and manual shutdown.

NOTE: Pressure maintenance pumps are covered later in this chapter.

Historically, pump controllers have used mercury switches to turn the pumps on and off **(Figure 4.13)**. In modern controllers, however, pressure sensing and control may be done through electronic means and be programmable through microprocessors with digital readout **(Figure 4.14)**.

Ways to measure current and voltage are indicated on the control panel. Measuring current and voltage is an integral part of pump testing, as will be discussed later in this chapter. Sometimes a pump is used to supply a deluge system or other fire protection system where it is necessary to turn the pump on independent of a pressure drop. In this case, relays can be used to turn the pump on when a signal is received from a fire- or smoke-detection device. If two or more pumps are connected in parallel, the controller should be capable, through sequential-timing devices, of preventing the motors from starting at the same time. If the demand becomes great enough that both pumps are needed, the controller is required to provide intervals of five to ten seconds between starting of additional pumps.

If the pump installation is the only source of water supply pressure for a sprinkler system or standpipe system, the automatic start controller must be wired for manual shutdown. This is done to ensure that the pump will be turned off *only* after a fire emergency is over. In addition, if the pump room is not constantly attended, audible or visible alarms should be transmitted to an attended location to signal that the pump is running or to signal that the power supply to the pump has been interrupted.

Some pumps may not be equipped with automatic controllers. If manual controllers are used, there will be both a manually operated electric switch to turn the pump on and off in addition to a mechanical control consisting of a lever or handle that can be latched in the ON position.

Diesel Engine Controllers

The controllers for diesel engines are not interchangeable with the controllers for electric motors. Electric motor controllers were designed to start the pumps by closing high-voltage circuits. On the diesel controller, the main function is to close the circuit between the storage batteries and the engine starter motor.

Two features to notice are the alarm and signal devices located on the controller itself. If the controller is automatic, a pilot light will indicate when the controller is in the AUTOMATIC posi-

Figure 4.13 Most fire pump controllers are activated by a mercury switch.

Figure 4.14 Newer fire pump controllers may be electronically activated.

tion. Separate lights and a common audible alarm are also required to indicate the following (**Figure 4.15**):

- Low engine oil pressure
- High engine coolant temperature
- Failure of engine to start automatically
- Engine shutdown due to overspeed
- Battery failure
- Low pressure in the storage tanks when air or hydraulic starting is used

The controller also requires separate lights for battery charger failure, but this does not require the provision of the audible alarm.

If the pump room is not constantly attended, alarm signals should be transmitted to a constantly attended location. The alarm will indicate when the engine has started, when the controller has been turned off or turned to manual operation, and when trouble exists with the engine or controller.

The controller may also be equipped with a pressure recording device. The recorder is usually required to run continuously for at least seven days without resetting or rewinding. The recorder's chart drive should be spring wound, AC electric with spring-driven backup, or air powered.

The provision for automatic starting of the engine due to water pressure or fire protection device activation is similar to the provision for the electric motor controller. If the pump installation is the

Figure 4.15 The diesel fire pump controller is designed to indicate a number of problems that require attention.

sole supply to standpipes or sprinkler systems, the automatic start controller should be wired for manual shutdown. In addition, the automatic controller usually must be arranged to automatically start the engine every week to ensure reliability in engine starting. If the controller is designed to automatically shut down the engine when system conditions return to normal, it should provide a running time of at least 30 minutes before shutting the engine off. However, if the overspeed governor operates, the controller is to shut off the engine without a time delay.

Pump Components and Accessories

The following additional components fit together to make a complete fire pump installation: pipe, fittings, power supply, relief valves, test equipment, pressure maintenance pumps, gauges, alarms, and the pump house. These are discussed further in the sections that follow. Also included is a discussion of the way these components are configured and standard performance specifications.

Pipe and Fittings

The aboveground pipe in a pump installation must be constructed of steel. The underground pipe may be of any material complying with the criteria set forth in NFPA 24, *Standard for the Installation of Private Fire Service Mains and Their Appurtenances*, 2002 edition. These other materials include plastic, asbestos cement, and ductile iron. The steel pipe can be joined together by screwed or flanged fittings or with grooved fittings. The pipe can also be welded together.

Suction pipe must be installed and tested in accordance with NFPA 24. Testing specifics are discussed later in this chapter. An OS&Y (Outside Stem and Yoke) type indicating control valve must be located in the suction line for control of the water supply. Butterfly valves are not permitted in the suction line due to the greater turbulence created by these types of valves, unless located at least 50 feet (16 m) away from the pump.

The discharge pipe should be hydrostatically tested in accordance with NFPA 13, *Standard for the Installation of Sprinkler Systems*, as well as NFPA 24.

A check valve is required in the discharge piping that permits water to flow in one direction only. This valve keeps water from flowing backward through the pump when the pump stops. An indicating control valve is also required in the discharge line. It is to be located on the system or downstream side of the check valve. This control valve may be either an OS&Y type or a butterfly type.

The proper sizing of both suction and discharge pipe is important. If piping size is too small, the pump cannot perform at its rated capacity. The size depends upon the rating of the pump. The information from the table in NFPA 20 is summarized in **Table 4.1**. Note that 5-inch (125 mm) pipe is the minimum acceptable size for a 500 gpm (2 000 L/min) pump. Because 5-inch (125 mm) pipe is not commonly available, 6-inch (150 mm) pipe is the smallest size likely to be found with any pump rated at 500 gpm (2 000 L/min) or larger.

In most cases, the minimum sizes are the same for both the suction side and the discharge side. For the 1,000 gpm (4 000 L/min) pump, the minimum for the suction side of the pump is larger than the discharge side. Although this is not a common requirement, in practice it is common to see the suction pipe larger than the discharge pipe. This is particularly true where a horizontal-shaft pump is receiving its supply from an aboveground tank. The larger piping reduces the friction loss and increases the pressure available at the pump.

When the size of the suction pipe is different from the inlet port of the pump, a reducer will need

Table 4.1 Minimum Pipe Size					
Pump Rating		Suction Pipe		Discharge Pipe	
(GPM)	(L/min)	(Inches)	(mm)	(Inches)	(mm)
500	2 000	5	125	5	125
750	3 000	6	150	6	150
1,000	4 000	8	200	6	150
1,250	5 000	8	200	8	200
1,500	6 000	8	200	8	200
2,000	8 000	10	250	10	250

to be installed in the suction line. The reducer is required to be of the eccentric type, not a concentric type (**Figure 4.16**). The reason that the reducer must be an eccentric type is to eliminate the possibility of air pockets. Air becoming entrained in the water stream can reduce the efficiency of the pump and can even damage it due to a phenomenon known as cavitation. The eccentric reducer is to be installed with the flat side on top.

NFPA 20 also states that no device that will restrict water flow can be installed on the suction side of a horizontal split-case pump. This includes backflow prevention devices, which are a common component on modern fire systems. Their proper location, therefore, would be on the discharge side of the pump. A strainer may be found on the suction side of a pump; it is most likely to be used where

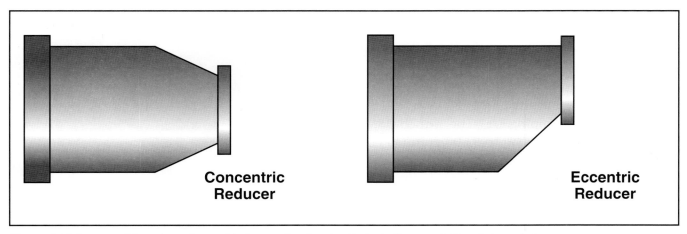

Concentric Reducer

Eccentric Reducer

Figure 4.16 An eccentric reducer is used in the suction line to eliminate the possibility of air pockets.

nonpotable (not suitable for drinking) water is supplied to a split-case pump from an underground tank. Strainers can also restrict the flow to the pump, especially if they are undersized or dirty.

Relief Valves

Historically, if a pump was driven by a variable-speed driver such as a diesel engine, a pressure relief valve was required in the installation. More recent versions of NFPA 20 require these large relief valves only if the pressure at churn is high enough to damage system components. The purpose of this relief valve is to prevent pressures that are high enough to damage system piping or fittings. It is possible for an internal combustion engine to get out of adjustment and develop excess speeds. Pump pressure is related to the square of the pump speed. For example, doubling the revolutions per minute increases the pressure developed by four times. Therefore, the relief valve is provided to open and discharge water to a drain if the pressure becomes excessive. The relief valve should be located between the pump and the discharge check valve. It is supposed to discharge to an open drain or in some manner so that water flow can be visually detected.

The size of the relief valve and its discharge line depend upon the rating of the pump. If the pump was chosen with a proper pressure rating, the relief valve should be set to open up a little above the normal discharge pressure when the pump is operating at churn.

Circulation Relief Valve

Although the large relief valve is ordinarily not present on electrically driven pump installations, a small relief valve will be present. This is called a circulation relief valve. It is designed to open and provide enough water flow into and out of the pump to prevent the pump from overheating when it is operating at churn against a closed system. This relief valve and the piping are to be ¾-inch (18.75 mm) in size for pumps up to 2,500 gpm (10 000 L/min) and 1 inch (25 mm) for larger pumps. All pumps are required to have this circulation relief valve, except for the diesel engine driven pump where engine-cooling water is taken from the pump discharge.

Set the circulation relief valve to flow a full stream when the pump is operating at churn, and then close off when water begins to flow in the system. Install the discharge piping to discharge into a drain where the flow from the line can be inspected.

Test Equipment

Every pump is required to be provided with components for testing the installation. The most prominent of these components is the test manifold **(Figure 4.17)**. The test piping should be connected to the pump discharge line between the check valve and the indicating control valve **(Figure 4.18)**. There should also be an indicating control valve in the test piping. This piping should terminate in a hose valve header located outside the building. The hose valve header should be equipped with 2½-inch (65 mm) hose connections with a shutoff valve for each connection **(Figure 4.19)**. This manifold and test header will permit water to be flowed from the pump installation through hoselines and nozzles for test purposes. The flow from the nozzles is measured using pitot tubes.

NOTE: NFPA 20 allows the use of metering devices as well as pitot tubes for testing. Consult the standard for more information.

Figure 4.17 A common pump test manifold.

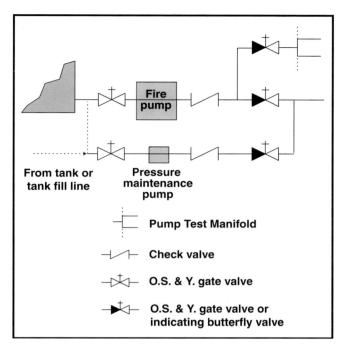

Pump Test Manifold

⊣⊢ **Check valve**

⊣⋈⊢ **O.S. & Y. gate valve**

▶⋈⊢ **O.S. & Y. gate valve or indicating butterfly valve**

Figure 4.18 This diagram shows the location of the test manifold on the pump system.

Figure 4.19 Each discharge is equipped with its own valve.

Table 4.2 Required Number of Test Valves				
Pump Rating		**Number of Hose Valves**	**Test Pipe Size**	
(GPM)	**(L/min)**		**(Inches)**	**(mm)**
500	2 000	2	4	100
750	3 000	3	6	150
1,000	4 000	4	6	150
1,250	5 000	6	8	200
1,500	6 000	6	8	200
2,000	8 000	6	8	200

Figure 4.20 Some pump installations are equipped with a metering device for measuring the gpm delivered by the pump.

The required size of the hose header supply pipe and the number of hose valves required depends upon the rating of the pump. It is often possible to estimate the rating of the pump by counting the hose connections. There is usually one 2½-inch hose (65 mm) connection for each 250 gpm (1 000 L/min) of pump rating. For example, a 500 gpm (2 000 L/min) pump usually has two hose connections and a 750 gpm (3 000 L/min) pump usually has three hose connections. As can be seen from **Table 4.2,** the method is not foolproof and breaks down with the 1,250 and 2,000 gpm (5 000 L/min and 8 000 L/min) pumps.

Some more recent pump installations do not have the test headers and hose valves; instead, they are equipped with a metering device that can be used to measure the gpm delivered by the pump **(Figure 4.20)**. This is acceptable, but the meter line should discharge to the outside or back to the water supply source, not directly back to the suction side of the pump. Circulating back to the suction side of the pump will not enable the condition of the suction supply to be evaluated.

Pressure Maintenance Pumps

Sometimes there is enough leakage in the fire protection system or enough fluctuation in the pressure of the water supply to the pump to cause the automatic controller to turn the pump on periodically in nonemergency situations. This can be a serious bother if it happens frequently, particularly in installations that require manual shutdown after automatic operation. A pressure maintenance pump is used to prevent these false starts.

A pressure maintenance pump is a small-capacity, high-pressure pump used to maintain constant pressures on the fire protection system. This pump takes suction from the fire pump suction line and discharges into the fire pump discharge line on the system side or downstream side of the indicating control valve. The pressure maintenance pump should have adequate capacity to keep up with any leaks in the system. It should be small enough, however, that any demand on the fire protection system, even a single sprinkler opening, will result in the main fire pump operating.

The pressure rating of the pressure maintenance pump should be high enough to maintain the desired fire protection system pressure. The pressure switch in the fire pump controller should be set to correspond to the system pressure maintained by the pressure maintenance pump. Thus, when the pressure maintenance pump cannot maintain the system pressure due to a demand on the system, the fire pump controller will activate the fire pump. The pressure maintenance pump installation requires the provision of a check valve in the discharge pipe from the pressure maintenance pump, as well as indicating control valves. In some cases, pressure relief valves are also required.

Pumps that are automatically controlled should be provided with an automatic air release. This air release is a float-type device that automatically closes off when the pump casing fills with water and allows the release of any air that may be trapped inside the pump. Remember that air inside the pump can reduce the pump efficiency and even damage the pump.

Gauges

Vertical turbine pump installations are required to have a single gauge. On a horizontal split-case pump and on vertical-shaft pump installations, there should be two gauges. One of the gauges should be located near the discharge port of the pump and the other near the intake. These gauges

should be at least 3½ inches (87.5 mm) in diameter and be capable of registering pressures of at least 200 psi (1 400 kPa) or twice the rated pressure of the pump, whichever is greater. Gauges of all sizes and with various pressure ranges are likely to be encountered, however. The suction gauge must be a compound gauge that registers both positive and negative gauge pressures. Often, an ordinary gauge will register only positive pressures.

A vertical turbine pump is required to have only a single gauge on the discharge line. Special attention should be given to these pressure gauges. Inaccurate gauges can invalidate test results and give no idea how well a pump is performing. Often during the testing of a pump, the gauge needle may vibrate so widely that an accurate reading is very difficult to obtain. If this is the case, the use of a liquid-filled gauge can eliminate the vibration problem.

Component Arrangement

The major components of a pump installation have now been identified. Shown in **Figure 4.21** is a line drawing of a typical fire pump installation using a horizontal split-case pump. Note that the indicating control valves on the supply sides of the fire pump and pressure maintenance pump must be of the OS&Y type. The other three control valves may be of the OS&Y or butterfly type. Note the locations of the check valves and the test header connection.

In some instances, the water supply for the pump may be taken from a city water supply system that has pressures high enough to be of value even without the pump. In this case, an additional bypass line should be provided around the pump (**Figure 4.22**). Ordinarily, this bypass line should be as large in diameter as the pump discharge line.

It may seem as though there are many indicating control valves required in a pump installation. The reason for so many valves is to be able to isolate any component, such as the pressure maintenance pump or the check valve, in the bypass line. This enables a component to be worked on or even removed without having to shut off the pump installation. **Figure 4.23, p. 140,** further illustrates a horizontal-shaft pump installation taking suction from an aboveground storage tank.

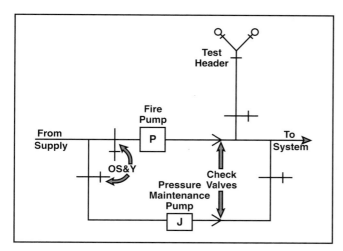

Figure 4.21 The standard arrangement for a fire pump installation.

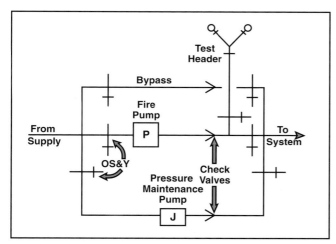

Figure 4.22 The arrangement for a fire pump installation that is equipped with a bypass.

The general piping arrangements and components of an installation using a vertical-shaft pump are shown in **Figure 4.24, p. 141**. Note that the components in the discharge side of the pump are essentially the same as they are for the horizontal pump. There is the relief valve, if needed, the check valve, and the test and discharge pipes with their indicating control valves. Notice the single pressure gauge in the discharge line.

A vertical-shaft pump often takes suction from a wet pit. This type of installation would ordinarily exist where the water supply source is a pond, lake, or river. An important feature here is the double screening required between the pond and pit that is designed to keep foreign matter out of the pump. The screens should be removable for easy cleaning.

Standard Performance Specifications

The primary performance criteria for standard fire pumps are contained in NFPA 20. The FM Global data sheets also contain information on fire pumps. For the most part, the Factory Mutual material comes from NFPA 20. In order to be considered standard under the provisions of NFPA 20, a new fire pump must be capable of satisfying three test points.

In the first test point, the pump must not develop more than 140 percent of its rated pressure when operating against a closed system (also referred to as churn or shutoff). However, it should be noted that all horizontal pumps manufactured before 1987 were required not to exceed 120 percent of the rated pressure at churn. These are maximum points.

Centrifugal pumps are designed so that the highest pressures are obtained at the lowest flow. Thus, when operating against a closed system with no fire protection water actually flowing, the highest pressures should be expected. When a pump is operating against a closed system, the pump is said to be operating at churn. Specifying a maximum acceptable increase in pressure at churn offers a degree of protection against excess pressures developing that could damage system components. So, if a pump is rated at 100 psi (700 kPa), no more than 140 psi (980 kPa) should be developed at churn.

NOTE: Additional information about test points for performance specifications can be found in **Appendix C, Pump Tests**.

Testing, Inspection, and Maintenance of Fire Pumps

Fire pumps that fail to operate when needed are likely to result in catastrophic losses. The way to prevent pump failure is to regularly ensure that pump installations are in good operating condition. The following sections look at the testing of new and existing installations in addition to inspection and maintenance of fire pumps.

It is recommended that pumps be operated weekly. Activate pumps from pressure drops and bring them up to full speed. It is not necessary to

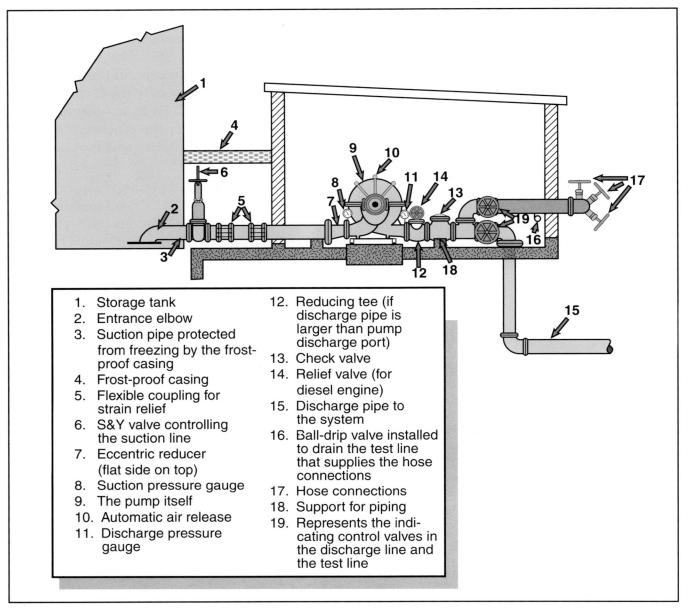

Figure 4.23 A horizontal split-case fire pump installation with water supply under a positive head.

1. Storage tank
2. Entrance elbow
3. Suction pipe protected from freezing by the frost-proof casing
4. Frost-proof casing
5. Flexible coupling for strain relief
6. S&Y valve controlling the suction line
7. Eccentric reducer (flat side on top)
8. Suction pressure gauge
9. The pump itself
10. Automatic air release
11. Discharge pressure gauge
12. Reducing tee (if discharge pipe is larger than pump discharge port)
13. Check valve
14. Relief valve (for diesel engine)
15. Discharge pipe to the system
16. Ball-drip valve installed to drain the test line that supplies the hose connections
17. Hose connections
18. Support for piping
19. Represents the indicating control valves in the discharge line and the test line

actually discharge water during the weekly start-up. Operate a diesel engine for at least 30 minutes and an electric motor for 10 minutes.

Testing Fire Pumps

Both the underground and aboveground piping must be hydrostatically tested in accordance with NFPA 24 and NFPA 13, respectively. Both types of piping must be hydrostatically pressurized for 2 hours to either 200 psi (1 400 kPa) or 50 psi (350 kPa) above the maximum static pressure, whichever is higher. For overhead piping, any leakage at all constitutes failure. The underground pipe is

permitted to leak a little, just a few quarts per hour, depending upon the length of pipe and the number and type of valves and fittings.

The underground pipe must also be flushed out prior to connection to the fire protection system piping. The reason for this is to flush out all of the debris accumulated in the pipe prior to installation. As might be expected, the required flow rate for flushing depends upon the diameter of the underground pipe. If the foreign materials are not flushed through the piping before connection to the fire protection system, these materials will end up inside the system piping. This can have

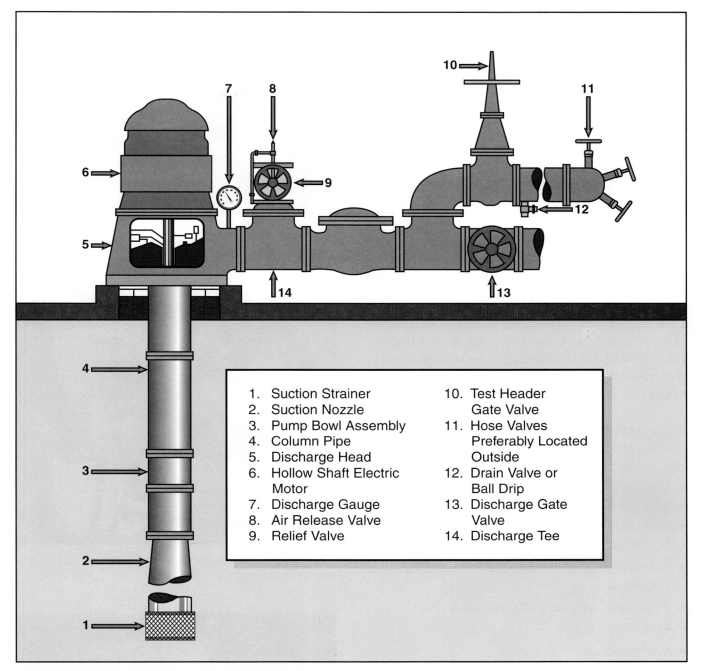

Figure 4.24 A vertical shaft, turbine-type fire pump.

1. Suction Strainer
2. Suction Nozzle
3. Pump Bowl Assembly
4. Column Pipe
5. Discharge Head
6. Hollow Shaft Electric Motor
7. Discharge Gauge
8. Air Release Valve
9. Relief Valve
10. Test Header Gate Valve
11. Hose Valves Preferably Located Outside
12. Drain Valve or Ball Drip
13. Discharge Gate Valve
14. Discharge Tee

a serious impact on the effectiveness of the fire protection system.

Before a pump installation is accepted from the installing contractor, the installation should be tested under the specifications of NFPA 20. Before the manufacturer ships a pump, it is tested in the shop. The results of this test will be plotted on graph paper. These plotted curves are called the certified shop test curves for the pump (**Figure 4.25, p. 142**). The features of performance that should be plotted are net pressure versus gpm delivered and horsepower delivered versus gpm delivered. These characteristic curves are the manufacturer's guarantee of the new pump's capabilities. An important requirement of the acceptance test is that the pump operates at least as well as the pump characteristic curves. The installation should not be accepted if it fails to meet the standard specifications.

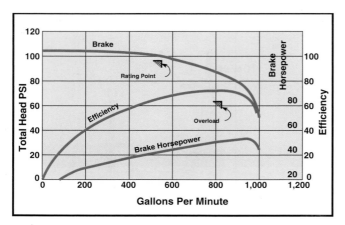

Figure 4.25 A typical fire pump shop curve.

Figure 4.26 Underwriters playpipes being prepared for the fire pump test.

The pump being tested must also meet the three standard performance points. At shutoff, not more than 140 percent of the rated net pressure may be developed. The pump must develop at least rated net pressure while delivering the rated flow and must develop at least 65 percent of the rated net pressure while delivering 150 percent of the rated flow.

Equipment Needed for Pump Tests

Besides the pump installation itself, some basic equipment is needed to conduct the test:

- One section of 2½-inch (65 mm) or larger hose for each hose connection on the test header

- One Underwriters playpipe nozzle for each hose-line (**Figure 4.26**)

- Method/device for safely securing playpipes (**Figure 4.27**)

- Pitot tube and gauge (**Figure 4.28**)

- Method of measuring pump speed

- Voltmeter

- Ammeter

If the system is equipped with a flow metering device, the first four items listed above are not necessary. At least one 50-foot (16.6 m) section of hose will be needed for each hose connection on the test header. This hose must have 2½-inch (65 mm) couplings with threads compatible with the hose connections in the header and be at least 2½ inches (62.5 mm) in diameter.

There should be an Underwriters playpipe for each hoseline. It is possible to use a deluge gun where several hoselines supply a single nozzle.

Figure 4.27 Some type of secure holder must be used to anchor the playpipes.

Figure 4.28 A pitot tube and gauge are used to measure flow (velocity) pressure.

However, the pressure lost in such devices may limit the flow capacity to less than could be obtained with separate playpipes. If a deluge gun is used, it must be equipped with a straight stream nozzle.

Figure 4.30 Insert the pitot tube into the stream to record the pressure.

Figure 4.29 Playpipes must be securely anchored for safety. In this case, the playpipe holder is anchored by having a car parked on it.

The purpose of the hoselines and playpipes is to allow measurement of the flow in gpm by means of pitot tube and gauge. Pitot tubes come in many shapes and sizes. A model with an air chamber for a handle or with a liquid-filled gauge reduces needle vibration and gives more accurate readings. For pump testing, gauges calibrated to at least 100 psi (700 kPa) are generally needed. The blade of the pitot tube is inserted into the flowing stream at a point one-half of the nozzle diameter away from the nozzle. The pressure registered on the gauge is the velocity pressure of the stream (**Figure 4.30**).

This velocity pressure is then mathematically converted to gpm.

Equation 4.1

$$Q = (29.83) (C_d) (D^2) (\sqrt{P})$$

Where: Q = Flow in gpm

C_d **= Coefficient of discharge**

D = Discharge orifice diameter

P = Nozzle pressure

Figure 4.31, p. 144, shows the common coefficient of discharges that might be used.

The rating of a pump is always for a specific pump speed. For example, a pump may be rated at 1,000 gpm and 60 psi (4 000 L/min and 420 kPa) at 1,770 revolutions per minute (rpm). The performance of such a pump will be certified at the speed of 1,770 rpm. If the electric motor or diesel engine cannot turn the pump this fast, the pump cannot be expected to meet the performance specifications. For this reason, some means of measuring the rpm must be available. This can be accomplished by using either a handheld revolution counter, a strobe-type tachometer, or by using the more modern digital tachometer. A voltmeter and an ammeter are needed for testing an electric motor driven pump.

Acceptance Test on Electric Pumps
The following are the 10 steps required to perform an acceptance test on an electric horizontal split-case pump:

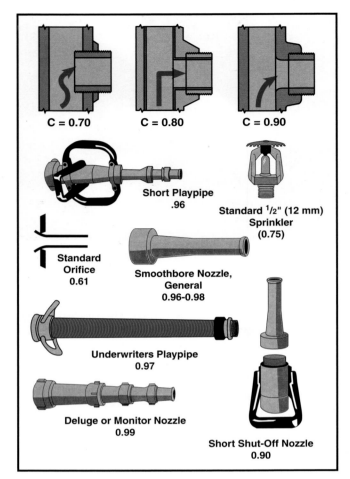

Figure 4.31 This chart shows the various coefficients for discharges that might be used for fire pump or hydrant testing.

Figure 4.32 Connect all the hoselines and nozzles.

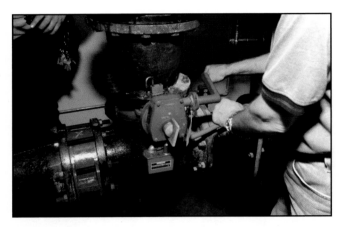

Figure 4.33 Close the valve that leads to the sprinkler/standpipe piping system.

Step 1: Calculate the expected pitot pressure for 100 percent and 150 percent of the rated flow by algebraically manipulating Equation 4.1 as follows:

$$P = \left(\frac{Q}{(29.83)(C_d)(D^2)} \right)^2$$

Step 2: Connect all the hoselines and nozzles. Make sure that all nozzles are securely fastened in place. Close all hose valves and the control valve in the pipe to the test header (**Figure 4.32**).

Step 3: Close off the indicating control valve that separates the pump from the fire system (**Figure 4.33**). This allows testing of the pump without subjecting the system piping to possible water hammer.

Step 4: Connect the ammeter and voltmeter to the test leads in the controller or at any other appropriate location.

CAUTION!
Only experienced personnel should work on the wiring of an electrically driven pump.

Step 5: If a handheld revolution counter is being used to measure pump speed, remove the end plate of the motor to gain access to the shaft. Because both the motor and the pump will be rotating at the same speed, measuring the speed of the motor also gives the speed of the pump.

Step 6: If the end of the shaft is not accessible, a strobe-type tachometer can be used. To establish the speed of the pump using the tachometer, mark the shaft with a piece of chalk and adjust the strobe impulse until the rotating chalk mark appears to be standing still. The pump speed can then be read from the tachometer dial. Several modern styles of digital tachometers are available. The most common variety uses a small strip of reflective tape that is placed on the shaft. The handheld sensor counts the rate of reflections as the shaft rotates.

Step 7: With everything ready, the pump can be started. Initially, the pump should be operating against a closed system with no valves open and with no water flowing. This is the churn or shutoff phase. The pump can be started manually or, if equipped to start automatically, should be started by bleeding off the water pressure. Once the pump is operating, read and record both the suction and discharge pressures, measure the rpm, and take voltage and current readings. While the pump is operating at churn, the circulation relief valve should have opened automatically and be flowing a solid stream of water. If water is not flowing, the adjust the relief valve with a crescent wrench until water begins to flow.

Step 8: Open the control valve in the line leading to the test header and open the hose valves for the first gpm measurement.

Step 9: Open and adjust sufficient lines so that the exact required pitot pressure for 100 percent of the rated flow is read on the pitot gauge (**Figure 4.34**). While the ve-

locity pressures are being measured outside, measure the rpm, voltage, current, discharge, and suction pressures inside the pump room. When the first line is opened, check the circulation relief valve again. When the first line was opened, the relief valve should have closed off. If it continues to flow a solid stream, further adjustment is needed.

Step 10: When all readings are complete and recorded, the additional hoselines are opened and adjusted to the exact required pitot pressures for 150 percent of the rated flow. This usually requires flowing all of the hoselines to achieve 150 percent of the rated flow. It is important to calculate ahead of time what pitot pressure is needed at each nozzle to give both the 100 percent and the 150 percent points. It is convenient if the total flows required are divided by the number of hoselines to be used to determine the flow required from each line. In this way, the pitot reading required at each nozzle will be identical.

In addition to the first phase, manually controlled pumps must be manually started and stopped at least six times with the pump running at least five minutes each time. An automatically controlled pump must be put through at least six automatic operations plus six manual operations with the pump running at least five minutes in each cycle. If the automatic controller is to start

Figure 4.34 Record the pressure for each hoseline.

the pump in response to a fire protection system operation, such as a fire detection system, this feature should be tested as well.

For the electric-driven pump, one start-up is to be conducted with all hoselines open to see if the pump will come up to rated speed under full load without pulling excess current and throwing the circuit breaker. All of these multiple operation tests are made to determine if the starting mechanism is operating properly. During all phases of the testing procedure, the pump is required to be in operation no less than one hour.

During the course of the test, pay attention to the temperature of the pump bearings and the pump itself. None of the components should become hot to the touch **(Figure 4.35)**. Fiber packing on both sides of the pump shaft seals the shaft; this packing is water-cooled and lubricated and some water will drip from the shaft at both ends. If the packing gland is adjusted too tightly, it will prevent water from cooling the fiber packing and the packing will heat up. It can become hot enough to burn and smoke. During the initial operations of the pump, gradual adjustment of the packing gland with a small wrench is often necessary. The goal is to achieve an adjustment where about one drop per second is passing through the packing. More leakage than this requires tightening the packing, while less leakage requires loosening.

When the test has been completed, the data collected is used to construct performance curves that are compared with the manufacturer's certi-

Figure 4.35 Feel the pump motor housing for signs of overheating.

fied curves. In constructing the pressure versus flow curve, it is the "net pressure" as discussed previously that is used. If the performance curve is falling very close to the characteristic curve, the velocity pressures should be considered. An increase or decrease in pipe size causes pressure changes because of the change in water velocity. But for most practical applications these pressure changes can be ignored.

Pump speed significantly affects the performance of the pump; this is the reason pump speed measurements must be taken. If the pump is turning at a speed other than the rated speed, any comparison with the certified performance may not be legitimate.

NOTE: A qualified technician should be called in to examine any pump that is not turning at rated speed. For more information about mathematically correcting pump speeds using affinity laws, consult the FPP *Fire Protection Hydraulics and Water Supply Analysis* manual.

The voltage and current measured for the electrically driven pump are used to evaluate other acceptance criteria. An electric motor should have a nameplate stamped with the service factor, full-load current rating, and rated voltage. During the acceptance test, the full-load current rating should not be exceeded except as allowed by the service factor.

The ratio of the measured current in amperes to the full-load current rating should not exceed the service factor any time during the test. In addition, the ratio of the products of the measured voltage times the measured current to the product of the rated voltage times the rated full-load current shall be less than or equal to the service factor.

Finally, the measured voltage should never be more than 5 percent below or more than 10 percent above the rated voltage.

For the vertical-shaft electrically driven pumps, the test procedure is essentially the same. The primary difference is that there is no suction gauge and the pressure developed is calculated by the procedure outlined in the first part of this chapter.

The test for diesel-driven pumps is the same as that for electrically driven pumps. Of course, voltage and current readings are not necessary, but attention should be given to engine temperature and oil pressure.

Regardless of the type of pump or driver, the actual performance of the pump is compared to the certified shop test curves provided by the manufacturer. If the pump does not meet or exceed the characteristic curves or should it malfunction in any way, the pump installation should not be accepted. The installing contractor in conjunction with the equipment manufacturer should be required to bring the installation up to standard.

Another consideration is the large volume of water that will be discharged during a test. A 1,000 gpm (4 000 L/min) pump operated for one hour can discharge more than 60,000 gallons (240,000 L) of water (**Figure 4.36**). Care should be taken to avoid erosion or property damage of any kind. If possible, it is better if the hoselines can discharge back into the water source to conserve water.

Figure 4.36 A great deal of water will be flowing from the test site. Take care to avoid property damage or injuries.

Routine Operation and Maintenance

NFPA 25, *Standard for the Inspection, Testing and Maintenance of Water-Based Fire Protection Systems*, 2002 edition, requires that an annual flow test of the pump assembly be performed to determine its ability to continue to attain satisfactory performance at shutoff, rated flow, and peak loads. By doing this testing, the performance of the pump can be compared year by year.

The manufacturer's recommendations for a preventive maintenance program must be implemented. If no maintenance program is available from the manufacturer, a program that adheres to the requirements outline in NFPA 25, 2002 Edition, Chapter 8 should be adopted. The maintenance program should incorporate a sequence of weekly, monthly, quarterly, and annual tests, along with recommended maintenance, that will ensure the continued satisfactory performance of the fire pump assemblies. Although some types of bearings do not require grease (namely, those that are rubber), special attention must be made to clean and lubricate all bearings. The appropriate quantity and the correct type of lubricant must be applied. On older fire pumps, it may be necessary to adjust the pump packing. These packings should be checked and adjusted every month. Some fire pumps may be equipped with mechanical seals rather than the fiber packings. If this is the case and the mechanical seals are in good shape, no leakage should be visible. In addition to the packing and seals, it is a good practice to recheck the pump alignment regularly. Misalignment of the pump drive couplings can cause excessive vibration and pump damage.

As pointed out previously, the primary purpose of the weekly test is to make sure that the pumps will start. Therefore, both diesel and electrically driven pumps should be started each week. NFPA 25 requires that pumps designed to turn on in response to drops in water pressure shall be started weekly by reducing the water pressure. This is often accomplished by using automatic timers.

During the weekly inspections, it is also important to make sure that the pump room is kept clean, dry, and free of combustible storage. The weekly inspection is a good time to ensure that all control valves that are supposed to be open are open and are supervised in the OPEN position by padlock and chain, electronic equipment, or tag and seal. This is also the appropriate time to make sure that water tanks and diesel fuel tanks are full.

During testing and inspection, personnel will eventually encounter a situation where the pump installation is not operating properly. NFPA 25, Annex C, provides a troubleshooting checklist that can help in identifying the causes of pump problems. For example, excessive leakage at the stuffing box could be the result of improper tightening, the packing being improperly installed, or it might be indicative of worn or scarred pump shafts. If the problem can be identified, several possible causes of the problem are suggested and possible corrective measures are outlined.

Fire pump installations are vital parts of a total fire protection system. Their proper selection, installation, and maintenance can be the difference between business as usual tomorrow and a catastrophe that could destroy people, property, and jobs.

Summary

- Pumps may be used to provide water supplies where existing supplies are deficient.

- Modern pumps must be able to meet the following three standard points:

  — A maximum of 140 percent of rated pressure at a flow of 0 gpm.

  — A minimum of 100 percent of rated pressure at 100 percent of rated flow.

  — A minimum of 65 percent of rated pressure at 150 percent of rated flow.

- To provide dependable service, fire pumps must be properly maintained and started weekly.

- A complete performance test should be conducted on each pump installation every year.

Standpipes and Hose Systems

1. Describe the basic components of a standpipe system.

2. Describe the different classes of standpipe systems and their intended uses.

3. Discuss the different types of standpipe systems and their advantages and disadvantages.

4. Describe the function of a fire department connection (FDC).

5. Describe water supply considerations for standpipes, hoselines, and sprinkler systems.

6. Discuss water pressure considerations for standpipes in high-rise buildings.

7. Describe the use of pressure-regulating devices.

8. Be able to list the main points to watch for during initial and in-service inspections of standpipes.

FESHE Objectives

Fire and Emergency Services Higher Education (FESHE) Objectives:
Fire Science Curriculum: Fire Protection Systems

• Comprehendtypes, components, and operation of automatic, special sprinkler systems, and standpipes.

Chapter 5
Standpipes and Hose Systems

Standpipe and hose systems provide a means for the manual application of water on fires in large, one-story buildings or in high-rise buildings. Horizontal standpipes are provided in large warehouses, factories, and shopping malls. Most buildings over four stories high are equipped with vertical standpipes. Many national and local codes address the topic of standpipes and hose systems. This chapter provides an overview of standpipes and hose systems.

The value of standpipes in expansive one-story structures is primarily one of expediency. Horizontal standpipes expedite fire control by reducing the manual effort and time needed to advance a hoseline several hundred feet to reach the seat of a fire. Horizontal standpipes can also facilitate the overhaul of fires that have been controlled by sprinkler systems by reducing the amount of hose needed to reach the area. In many high-rise buildings, a standpipe is the primary means for *manually* controlling a fire and is an essential aspect of the building's design. Because the fire pump discharge pressure necessary to reach the top floor of a 50-story building may be 350-400 psi (2 450 kPa to 2 800 kPa), the proper operation of the standpipe system is critical.

A standpipe system can be a very simple system consisting only of a vertical pipe, called a *riser*, with hose connections and a fire department connection. A standpipe system can also be a very complex building system consisting of multiple pumps and risers, valves, and reservoirs located on the upper floors **(Figure 5.1)**. Standpipes can be designed to supply small-diameter hose for use by building occupants, to supply fire department hose, or both. Although standpipe systems are a necessity in high-rise buildings, they neither take the place

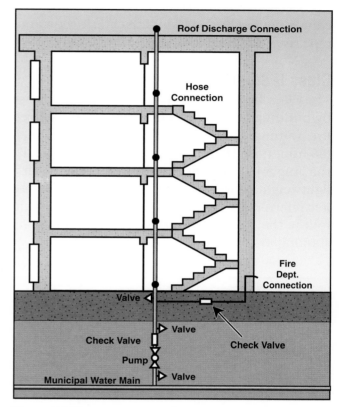

Figure 5.1 The components of a typical standpipe system.

of nor lessen the need for automatic sprinkler systems. Automatic sprinklers are still the most effective method of fire control in high-rise or other hazardous occupancies.

NOTE: Automatic sprinklers are discussed in more detail in Chapter 6 of this manual.

Classification of Standpipe Systems

NFPA 14, *Standard For the Installation of Standpipe and Hose Systems*, 2000 Edition, is frequently used for the design and installation of standpipes. The standard recognizes three classes of standpipe systems: Class I, Class II, and Class III. Each type is discussed in the sections that follow.

Class I Standpipe Systems

Class I standpipe systems are primarily for use by fire fighting personnel trained in handling large handlines (2½-inch [65mm] hose). Class I systems must be capable of supplying effective fire streams during the more advanced stages of fire within a building or for fighting a fire in an adjacent building. Class I systems have 2½-inch (65 mm) hose connections or hose stations attached to the standpipe riser (**Figure 5.2**).

Class II Standpipe Systems

The Class II system is primarily designed for use by building occupants who have no specialized fire training. These systems are limited to 1½-inch (38 mm) hose (**Figure 5.3**). This hose is typically the single-jacket variety and is equipped with a lightweight, twist-type shut-off nozzle. Sometimes it is referred to as a *house line*. Firefighters should follow their department SOPs when operating standpipes.

Class III Standpipe Systems

Class III standpipes combine the features of Class I and Class II systems. Class III systems have both 2½-inch (65 mm) connections for fire department personnel and 1½-inch (38 mm) hose stations and connections for use by building occupants (**Figure 5.4**). The design of the system must allow both the Class I and Class II services to be used simultaneously.

Figure 5.2 This Class I standpipe connection is designed to be used by firefighters, so features a larger hose connection.

Figure 5.3 Class II standpipe hose is designed for use by building occupants who have had little if any training in handling hose.

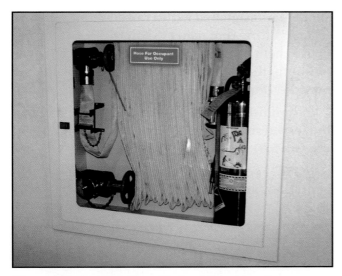

Figure 5.4 Note that Class III standpipes have the smaller hose connection and the larger 2 ½-inch (65 mm) connection for use by firefighters.

Types of Standpipe Systems

In addition to the classes of standpipes, the different types of standpipe systems are as follows:

- An automatic wet standpipe system is maintained wet at all times. The water supply is capable of supplying the system demand automatically. When a hose valve is opened, water is immediately available. A wet standpipe with an automatic water supply is most desirable because water is constantly available at the hose station. Wet standpipe systems cannot be used in cold environments; therefore, a dry system may have to be used.

- An automatic dry standpipe system is a dry standpipe normally, with air under pressure arranged to admit water into the system through a dry pipe valve upon the opening of a hose valve. Automatic dry systems have the disadvantages of greater cost and maintenance requirements. The main advantage of a dry standpipe with no permanent water supply is the reduction in cost.

- A semiautomatic dry system standpipe is a dry standpipe system that is arranged to admit water into the system when a remote control device is activated at the hose stations.

- A manual dry standpipe is a system that does not have a permanent water supply and is supplied only through a fire department connection.

- A manual wet system is a system that is maintained full of water from a small source that is incapable of supplying the system demand. A fire department connection must be used to supply system demand.

A primed system is a wet standpipe system that has a limited water supply that keeps the system riser full. Adequate water for fire fighting is not available until the fire department connects to the fire department connection or until a manual fire pump is started. The advantages of a primed system are as follows:

— Reduction of time for water to reach the hose valves

— Reduction in water hammer

— Reduction in corrosive effects on the internal surfaces of the standpipe

Fire Department Connections

Each Class I or Class III standpipe system requires one or more fire department connections through which a fire department pumper can supply water **(Figure 5.5)**. High-rise buildings having two or more zones require a fire department connection for each zone. In high-rise buildings with multiple zones, the upper zones may be beyond the height to which a fire department pumper can effectively supply water. This height would be around 450 feet (150 m) for a two-stage pumper, depending on available hydrant pressure and other factors. For standpipe system zones beyond that height,

Figure 5.5 Fire department connections (FDCs) enable fire pumps to supply water to standpipe systems.

a fire department connection is of no value unless the fire department is equipped with special high-pressure pumpers and the system has high-pressure piping. Standard requirements specify that there be no shutoff valve between the fire department connection and the standpipe riser. In multiple-riser systems, however, gate valves are provided at the base of the individual risers.

A fire department connection, or FDC, is a connection through which the fire department can pump supplemental water into a standpipe system or a sprinkler system. The FDC is located in an accessible area around the exterior of the building. A hose connection is a pipe and fitting for the connection of a hose to the standpipe system. It includes a hose valve and a threaded outlet. The hose connection is located within the interior of a building in a hose station, hose cabinet, in the stairwell, and sometimes on the roof. To simplify, water is pumped into a fire department connection (FDC) and out of a hose connection.

The hose connections to the fire department connection must be female and equipped with standard caps (**Figure 5.6**). The caps shown are designed to be brittle and easily broken from the connection if they are needed. Some jurisdictions require Storz-type (sexless) couplings that allow large-diameter hose to be used to supply standpipes. It is important that the hose coupling threads conform to those used by the local fire department. The fire department connection

Figure 5.6 The caps on FDC hose connections are designed to be broken easily.

should be designated with a raised-letter sign on a plate or fitting that reads "STANDPIPE." If the fire department connection does not service the entire building, the sign must indicate which floors are serviced (**Figure 5.7**).

In building complexes, the address or name of the served structure should be labeled on the FDC (**Figure 5.8**).

NOTE: For more information on fire department operations at standpipes, see **Appendix D**.

Figure 5.7 A zoned fire department connection must clearly indicate which floors it services.

Figure 5.8 Fire department connections that service building complexes must clearly indicate the address or name of the building.

Water Supply Considerations

The water supply for standpipes may come from such different sources as public water supplies, pressure tanks, and gravity tanks **(Figure 5.9)**. Not all of these water sources are practical in every situation. In high-rise buildings, the water is usually supplied from municipal water mains and pressurized through automatic or manual fire pumps located inside the building or complex. Water supplies can be used in combination; for example, it is possible to incorporate both a tank supply and an automatic fire pump in the supply for a high-rise building.

Standpipes

The amount of water required for standpipe systems depends on the size and number of fire streams that are needed and the probable length of time the standpipe will be used. These factors are influenced by the size and occupancy of the building and should be in accordance with state or local codes. The water supply for Class I and Class III standpipe systems should provide 500 gpm (2 000 L/min) for at least 30 minutes, with a residual pressure of 100 psi (700 kPa) at the most hydraulically remote 2½-inch (65 mm) outlet. A minimum of 100 psi (700 kPa) is required for the most remote 1½-inch (38 mm) outlet. If more than one standpipe riser is needed to protect a building, the water supply must provide 250 gpm (1 000 L/min) for each additional riser to a maximum of 2,500 gpm (10 000 L/min). For a Class II standpipe, 100 gpm (400 L/min) must be provided for at least 30 minutes, with a residual pressure of at least 65 psi (455 kPa) at the highest outlet.

Standpipes in High-Rise Buildings

The size of the standpipe riser is determined by the height of the building and the class of service. For Class I and Class III service, the minimum riser is 4 inches (100 mm) for building heights less than 100 feet (30 m) and 6 inches (150 mm) for heights over 100 feet (30 m). When a Class I or Class III standpipe exceeds 100 feet (30 m) in height, the top 100 feet (30 m) is allowed to be 4-inch (100 mm) pipe. Standpipes that are part of a combined system (sprinkler and standpipe) are to be 6 inches (150 mm). Standpipes can also be sized hydraulically

Figure 5.9 Gravity tanks can serve as supply sources for standpipes.

to provide the minimum required pressure at the topmost outlet. For Class II service, a riser could be 2 inches (50 mm) for a building height less than 50 feet (15 m). For a building over 50 feet (15 m) in height, the minimum size riser is 2½ inches (65 mm). Class II systems in buildings over 275 feet (91 m) in height should be divided into sections. In buildings with combined standpipe and sprinkler systems, the minimum riser size is 6 inches (150 mm). However, depending on local codes, this requirement may be disregarded if the building is completely sprinklered and the system is hydraulically calculated to ensure that all water supply requirements can be met.

Current practice for nonsprinklered buildings is to locate standpipes so that any part of a floor is within 150 feet (50 m) of the standpipe hose connection. This allows any fire to be reached with 120 feet (40 m) of hose, plus a 30-foot (10 m) fire stream. For sprinklered buildings, the travel distance is increased to 200 feet (66 m). Standpipes and their connections are most commonly located within noncombustible fire-rated stair enclosures so that firefighters have a protected point from which to begin an attack **(Figure 5.10)**. If the building is so large that the standpipes located in the stairwells cannot provide coverage to the entire floor, additional stations or risers should be provided. The actual hose connections can be located no more than 3 to 5 feet (1 to 1.2 m) from floor level. These connections should be plainly visible and should not be obstructed. Caps over the connections should be easy to remove.

Buildings equipped with Class I or Class III systems may be required to have a 2½-inch (65 mm) outlet on the roof. This outlet may be required when any of the following situations are present:

- The building has a combustible roof.
- The building has a combustible structure or equipment on the roof.
- The building has exposures that present a fire hazard.

Hoselines

The current NFPA 14 requirement for a residual pressure of 100 psi (700 kPa) is a minimum pressure and may not be adequate to supply a fog nozzle on the end of a 100-foot (30 m) hose connected to the topmost hose outlet. Because of this problem, some other building codes and fire codes require higher minimum residual pressures. For example, the Uniform Building Code requires 100 psi (700 kPa) for a residual pressure.

NOTE: Firefighters should consult the code or codes used in their jurisdictions for minimum requirements.

Sprinkler Systems

In addition to supplying water for hose streams, many standpipe risers are also used to supply water for the sprinkler system in high-rise buildings **(Figure 5.11)**. In a fully sprinklered building, it is expected that the sprinklers will extinguish or control an incipient fire and thereby reduce the water required for hoseline operations. Therefore, when determining the required water supply for a standpipe system in a fully sprinklered building, it

Figure 5.10 This stairwell gives firefighters some protection when they are fighting fire in a building.

Figure 5.11 A combination riser used to supply both a sprinkler system and standpipe hose valves.

is not necessary to add the sprinkler water demand to the water supply requirements for the standpipe. If, however, the sprinkler water demand is greater than the standpipe demand, the water supply must be adequate to meet the greater requirements of the sprinkler system.

Water Pressure Considerations

One unavoidable problem with standpipes in high-rise buildings is the increase in water pressure requirements because of the height of the building. It is important to remember that 0.434 psi is required to raise water 1 foot (10 kPa per meter). For example, a standpipe serving a 20-story building may be 200 feet (66 m) high, and 100 psi (700 kPa) must be provided to the top of the standpipe, per NFPA 14 **(Figure 5.12)**. The pressure required to overcome the elevation loss would be:

200 feet × 0.434 psi/foot = 87 psi

(60 meters × 10 kPa/meter = 600 kPa)

87 psi + 100 psi = 187 psi

(600 kPa + 690 kPa = 1 290 kPa)

Thus, the pressure at the base of the standpipe would need to be at least 87 psi (600 kPa) just to overcome the elevation change. If, however, the building were 400 feet (133 m) high, the pressure at the base would need to be at least 174 psi (1 200 kPa), which does not include friction loss.

400 feet × 0.434 psi/foot = 174 psi

(120 meters × 10 kPa/meter = 1 200 kPa)

174 psi + 100 psi = 274 psi

(1 200 kPa + 690 kPa = 1 890 kPa)

This high pressure at the lower floors is difficult to handle when hoses are attached to the lower hose valves. To avoid extremely high pressure at the lower hose stations, standpipes are divided into zones when the building height exceeds 275 feet (91 m) unless pressure-reducing valves are used at the hose stations. If pressure-reducing valves are used, the height of a zone may be extended to 400 feet (133 m). High rises may have several zones. The Sears Tower in Chicago, for example, is 1,400 feet (426.7 m) tall and has a standpipe system divided into 7 zones. The water for the upper zones is supplied by an individual *zone fire pump* that

Figure 5.12 A pressure of 100 psi (700 kPa) is still needed at the top of a standpipe in a high-rise building.

is supplied directly from a water main, or it may intake from the fire pump that supplies the lower zone. When one pump intakes from the discharge of another, the pumps are said to be arranged in *series*.

In high-rise buildings with several zones, the upper zone pumps may be arranged to draft from tanks on the upper floors. The tanks, which hold several thousand gallons (liters) of water, are filled automatically from lower zone fire pumps and/or the domestic water supply pumps by means of automatic float valves. These float valves open when the level in the tank begins to drop. Tanks located on the upper floors for a source of supply provide for greater system reliability. Should the lower zone pumps fail, there is still water available to the pumps located on the upper floors of the building.

Pressure-Regulating Devices

Where the discharge pressure at a hose outlet exceeds 100 psi (700 kPa), NFPA 14 requires a pressure-regulating, or restricting, device to limit the pressure to 100 psi (700 kPa), unless otherwise approved by the fire department. The use of a pressure-regulating device prevents pressures that make hose difficult or dangerous to handle. This device also enhances system reliability because it extends individual zones to greater heights. In some instances, it may improve system economy because its use may eliminate some pumps. However, pressure-regulating devices make the system design more complex.

There are several different types of pressure-regulating devices. One type consists of a simple restricting orifice inserted in the waterway. The pressure drop through the orifice plate depends on the orifice diameter and the flow. The individual restricting orifice should be sized for different applications and will not be the same for each floor of a given building.

Another type of pressure-regulating device may consist of vanes in the waterway that can be rotated to change the cross-sectional area through which the water flows.

A pressure-regulating device may also take the form of a pressure-reducing valve (**Figures 5.13 a and b**). There are several different pressure-reducing valves available from various manufacturers. Some of the valves are field adjustable and others are set at the factory.

A pressure-regulating device should be specified and/or adjusted to meet the pressure and flow requirements of the individual installation. For factory-set devices, the pressure-regulating device should be installed on the proper hose outlet to ensure proper installation. When field-adjustable devices are installed, the manufacturer's instructions on making adjustments must be followed carefully. If a pressure-regulating device is not properly installed or is not properly adjusted for the required inlet pressure, outlet pressure, and flow, the available flow may be greatly reduced and fire fighting capabilities seriously impaired.

Figures 5.13 a and b A pressure-reducing valve and a cutaway view of the same valve.

Inspecting and Testing Standpipes

In order to ensure both compliance with local codes adopted by the jurisdiction and standpipe operability, standpipes should be inspected when they are first installed and periodically thereafter. The following sections highlight inspection and testing procedures for standpipes. In addition to NFPA standards, check local codes and ordinances regarding standpipe installations.

Initial Installation Inspection and Tests

A standpipe system is a significant component in a building's design. Before system installation, detailed design plans should be submitted to the local fire department or building department. The plans are then checked for compliance with current codes that have been adopted by the jurisdiction. As construction proceeds, the installation should be checked for conformity with the plans. In a high-rise building, it will be necessary to have the standpipe in partial operation as construction proceeds. It will provide protection to the structure should fire occur on the upper levels during construction.

When the installation is complete, the following test and inspections should be performed:

- The system should be hydrostatically tested at a pressure of at least 200 psi (1 400 kPa) for two hours to ensure tightness and integrity of fittings. If the normal operating pressure is greater than 150 psi (1 050 kPa), the system should be tested at 50 psi (350 kPa) greater than its normal pressure.

- The system should be flow tested to remove any construction debris and to ensure that there are no obstructions.

- On systems equipped with an automatic fire pump, a flow test should be performed at the highest outlet to ensure that the fire pump will start when the hose valve is opened.

- The fire pump should be tested to ensure that it will deliver its rated flow and pressure.

- All devices should be inspected to ensure that they are listed by a nationally recognized testing laboratory such as Underwriters Laboratories or FM Global.

- Hose stations and connections should be checked to ensure that they are in cases within 5 feet (1.2 m) from the floor and are positioned so that the hose can be attached to the valve without kinking.

- Each hose cabinet or closet should be inspected for a conspicuous sign that reads "FIRE HOSE" and/or "FIRE HOSE FOR USE BY OCCUPANTS OF BUILDING" **(Figure 5.14)**.

- Fire department connections should be checked for the proper fire department thread and for a sign indicating "STANDPIPE" with a list of the floors served by that connection.

- When a dry standpipe is installed, check for a sign indicating "DRY STANDPIPE FOR FIRE DEPARTMENT USE ONLY."

Figure 5.14 Standpipes need to be clearly marked.

In-Service Inspections

As with all fire protection systems, standpipe systems need to be inspected and tested at regular intervals. A visual inspection should be made at least monthly by the building management. Because interior fire fighting is dependent on the standpipe system, the fire department should also inspect standpipes at regular intervals. Actual testing of standpipes should be carried out by a fire protection contractor or the building operating staff (if they are sufficiently knowledgeable). To avoid potential liability, it is prudent that fire department personnel do not perform actual tests. However, fire department personnel should witness tests.

Fire department personnel should reinspect standpipe systems for the following:

- All water supply valves are sealed in the open position.
- Power is available to the fire pump.
- Individual hose valves are free of paint, corrosion, and other impediments.
- Hose valve threads are not damaged and match fire department couplings (**Figure 5.15**).
- Fire department connection caps are in place. If not, replacements need to be requested.

Figure 5.15 It is important to verify that hose valve threads are in good condition.

- Pipes are free of trash or debris.
- Hose valve wheels are present and not damaged.
- Hose cabinets are accessible.
- Hose is in good condition, has proper dryness, and is properly positioned on the rack.
- Hose nozzles are present and in good working order.
- Discharge outlets in a dry system are closed.
- Dry standpipe is drained of moisture.
- Access to the fire department connection is not blocked.
- The fire department connection is free of obstruction and the swivels rotate freely.
- Water supply tanks are at the proper level.
- Any pressure-regulating devices are tested as required by the manufacturer.
- Hose valves on dry systems are closed.
- Dry systems are hydrostatically tested every five years.

Summary

Standpipes and hose systems reduce the time and effort needed by fire fighting personnel to attach hoselines and apply water during a fire emergency. In large or very tall buildings, much more effort and resources would be required to stretch hoselines if standpipes or fire department connections were not already in place. Because every minute counts during an emergency, the availability of these systems plays a crucial role in timely mitigation. Because it is easy to take built-in systems for granted, fire fighting personnel must see that they are inspected and in good working order. There is no time to repair a system during an emergency.

Automatic Sprinkler Systems

1. Name three reasons why sprinkler systems are installed.

2. Name the basic components of an automatic sprinkler system.

3. Identify the three main components of a sprinkler.

4. Describe the function of a heat-sensitive device.

5. Name the two major types of heat-sensitive devices.

6. Describe the ways in which temperature ratings of sprinklers can affect their performance.

7. Describe the way in which sprinkler design affects its ability to control a fire.

8. Name the basic types of sprinkler design and their uses.

9. Discuss the theory and function of Early-Suppression Fast-Response (ESFR) sprinklers.

10. Name the basic types of piping used in sprinkler systems.

11. Describe the function of the water supply main.

12. Describe system risers and their uses.

13. Describe the function of the main control valve.

14. Describe the function of a check valve.

15. Describe the main function of a fire department connection (FDC).

16. Name the four main types of sprinkler systems.

17. Describe the function of a retard chamber.

18. Describe the operating action of a dry-pipe sprinkler.

19. Describe the purpose of a quick-opening device.

20. Describe the way in which a deluge sprinkler system operates.

21. Name three ways in which deluge valves can operate.

22. Describe the way in which a preaction sprinkler system operates.

23. Describe ways in which warehouses present special extinguishing challenges.

24. Name the classes of commodities and their characteristics.

25. Name the reasons why fire inspection personnel should not manipulate machinery or attempt to perform maintenance.

26. Describe what to look for when inspecting sprinklers.

27. Discuss the ways in which changes in occupancy affect sprinkler design and usefulness.

28. Name the main areas that should be inspected in a wet-pipe sprinkler system.

29. Name the main differences between inspecting wet-pipe systems and dry-pipe systems.

30. Name the best way to ensure that a proper replacement sprinkler is selected.

31. Name some additional fire safety measures that can be applied when a sprinkler system is out of service.

32. Name the main operating differences between residential sprinklers and conventional sprinklers.

**Fire and Emergency Services Higher Education (FESHE) Objectives:
Fire Science Curriculum: Fire Protection Systems**

- Comprehend types, components, and operation of automatic, special sprinkler systems, and standpipes.

- Identify and describe appropriate national standards governing the installation, inspection, and maintenance of given extinguishing agent/systems and their related components.

Chapter 6
Automatic Sprinkler Systems

In their basic form, automatic sprinkler systems have been in use for over 100 years. Their origin dates back to the large industrial mills that dotted the northeastern United States during the Industrial Revolution. An American named Henry S. Parmalee, who was looking for a way to protect his piano factory, invented the first sprinkler system in 1874.

Today, the automatic sprinkler system is an unsurpassed fire protection device. Fire loss data reveal that automatic sprinklers have a 96 percent reliability factor when installed and maintained properly. In 29 percent of these fires, one sprinkler controlled the fire. In 75 percent of the reported fires, ten or fewer sprinklers controlled the fire. Of the fires that were not controlled in sprinkler-equipped buildings, failure was due to improper maintenance, an inadequate or inoperative water supply, incorrect design, obstructions, or partial protection.

Modern sprinkler system technology is both highly effective and reliable when designed, installed, and maintained properly. Today sprinkler systems are installed in schools, health care facilities, high-rise buildings, commercial buildings, residences, and many other types of occupancies. The installation, maintenance, and testing of sprinkler systems must be in compliance with state, local, and/or federal codes and ordinances. Many installations also follow applicable National Fire Protection Association, Underwriters Laboratories, FM Global, or insurance company guidelines.

NOTE: The material in this chapter is intended to be informative and descriptive, but it should not be used as the authority over local codes, ordinances, and standards.

Fires that involve large unsprinklered properties pose a threat to the entire community and overtax fire fighting resources **(Figure 6.1)**. Building owners invest in and install automatic sprinklers for the following reasons:

Figure 6.1
Any fire that is uncontrolled has the potential to take lives and damage surrounding properties. *Courtesy of District Chief Chris E. Mickal.*

- Code requirements
- Insurance incentives or requirements
- General fire protection of life and property
- Building design flexibility
- Inherent risk

Building and fire codes frequently require installation of automatic sprinklers. The reasons for mandating sprinklers in buildings arise from a need to protect the community as a whole or to protect the occupants in individual buildings of high occupancy design. Model codes require the installation of automatic sprinklers in buildings based on their occupancy, construction type, and size. Typically, when a building exceeds a given space limitation established in a building code, it is required to have sprinklers.

Automatic sprinklers may be installed in the interest of general fire safety even where codes do not require automatic sprinklers and insurance is not a factor. When an investment of several million dollars has been made in a business, good risk management suggests that the investment be protected from destruction by fire. In view of the prominent role sprinkler systems play in fire protection, an understanding of these systems and their components is essential to fire department personnel (**Figure 6.2**). It is especially important to understand sprinkler system operation so that fire fighting operational procedures can be carried out more efficiently in buildings protected with sprinklers.

The purpose of this chapter is to cover the various types of automatic sprinkler systems. Procedures for inspecting, testing, and resetting systems are highlighted. Fire department operations at sprinklered occupancies are also addressed in **Appendix E, Fire Department Operations at Sprinklered Occupancies**.

Components of Sprinkler Systems

The ideal fire control system should be simple, reliable, and automatic. It should use a readily available and inexpensive extinguishing agent and should discharge the extinguishing agent directly on the fire while it is in its incipient stage. The sys-

Figure 6.2 Familiarity with an occupancy's sprinkler system is a vital part of pre-incident planning. *Courtesy of Ted Boothroyd.*

tem that most closely meets those requirements is the automatic fire sprinkler system. According to the National Fire Sprinkler Association, there has never been a multiple loss of life (three or more deaths) due to fire or smoke in a building protected with a fire sprinkler system, except in the case of intimate contact with the fire or an explosion. This is a remarkable statistic considering how long sprinkler systems have been in use.

The fundamental concept of the automatic sprinkler system is very simple (**Figure 6.3**). Basically, an automatic sprinkler system consists of the following:

- Suitable water supply
- System of distribution piping
- Number of nozzles called sprinklers
- Number of valves and trim

Standards Related to Automatic Sprinkler Systems

Throughout this chapter, various National Fire Protection Association (NFPA) standards related to automatic sprinkler systems are referenced. This section lists those standards and gives a brief synopsis of each.

NFPA 13, *Standard for the Installation of Sprinkler Systems* (2002 edition)

NFPA 13 provides the minimum requirements for the design and installation of all types of sprinkler systems that provide a reasonable degree of protection for life and property. The only occupancies not specifically covered in NFPA 13 are smaller residential occupancies. This standard covers all aspects of system design including components, water supply, and fire pumps.

NFPA 13D, *Standard for the Installation of Sprinkler Systems in One- and Two-Family Dwellings and Mobile Homes* (2002 edition)

This standard lists the requirements for small, fast-response sprinkler systems that increase the life safety factor in one- and two-family private dwellings, including manufactured homes.

NFPA 13R, *Standard for the Installation of Sprinkler Systems in Residential Occupancies up to and Including Four Stories in Height* (2002 edition)

This standard provides requirements for residential-type sprinkler systems in residential occupancies up to and including four stories in height.

NFPA 25, *Standard for the Inspection, Testing, and Maintenance of Water-Based Fire Protection Systems* (2002 edition)

In addition to other systems, this standard provides the minimum requirements for the periodic inspection, testing, and maintenance of automatic sprinkler systems. The standard also addresses impairment handling and reporting. NFPA 25 does not cover inspection, testing, and maintenance of sprinkler systems as described in NFPA 13D.

NFPA 230, *Standard for the Fire Protection of Storage* (2003 edition)

This standard provides requirements for sprinkler system coverage in occupancies that utilize high piled or other unusual storage methods.

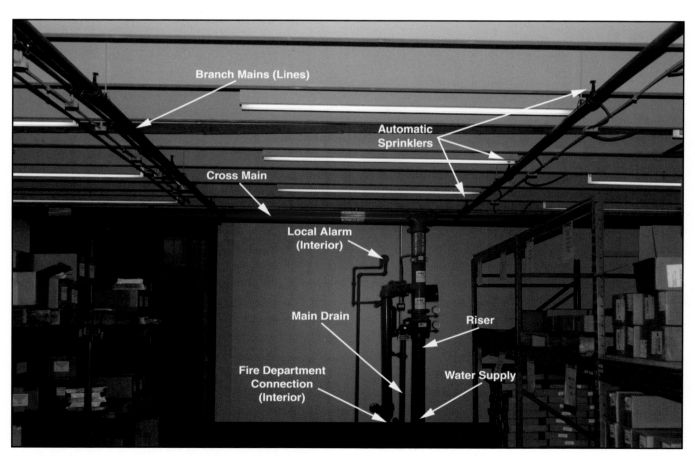

Figure 6.3 Components of a complete sprinkler system.

The sprinkler nozzles are arranged so that the system will automatically distribute sufficient quantities of water to either extinguish or control a fire and prevent flashover until firefighters arrive. Sprinkler systems are designed to provide protection in a wide variety of specific situations. Although they may become fairly sophisticated in actual application, the fundamental concept is still simple. It is the simplicity of the automatic sprinkler system that gives rise to its greatest virtue: reliability.

NOTE: See **Appendix F, Design Considerations of Automatic Sprinker Systems** for more information about sprinkler design.

Sprinklers

The *sprinkler*, also known as a sprinkler head, is that portion of the sprinkler system that senses the fire, reacts to that sense, and then delivers the water to the fire area. There are three main components of a sprinkler that are of interest to firefighters: the heat-sensitive device, the deflector, and the discharge orifice. The following sections highlight each of these components.

NOTE: Some systems may have open sprinklers or nozzles that do not have a heat-sensitive element.

Heat-Sensitive Device

A simple heat-sensitive device controls most automatic sprinklers. One of the most common types of these devices is the fusible link **(Figure 6.4)**. In its simplest form, the fusible link is a solder link with a low, precisely established melting point. The solder link is connected to a cap that restrains the water at the nozzle orifice. As illustrated in **Figure 6.5**, the solder in the fusible link melts at its predetermined melting point. The lever arms or struts are released and spring clear of the sprinkler frame. As the lever arms drop, the seated cap is released, which permits the water to flow **(Figure 6.6)**.

Another type of heat-sensitive device is the frangible (glass) bulb **(Figure 6.7)**. Frangible bulbs are inserted between the sprinkler frame and the discharge orifice in much the same way as the fusible link. Frangible bulbs contain liquids that expand when heated. The expansion of the liquid within the bulb increases vapor pressure. Once a vapor pressure is reached that exceeds the strength of the bulb, the bulb breaks, allowing water to flow.

Temperature Ratings

The components of a sprinkler can be varied to fit specific applications. By changing the composition of the solder, the operating temperature of the

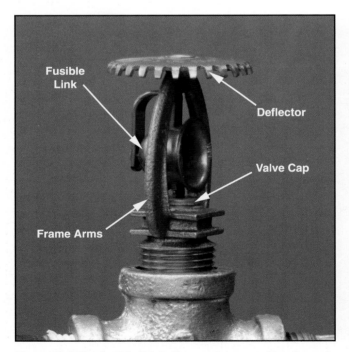

Figure 6.4 The fusible link is the most common heat-sensitive device installed in an automatic sprinkler system.

Figure 6.5 As the solder in the fusible link melts, the lever arms or struts are released.

Figure 6.6 As the seated cap is released, water begins to flow from the sprinkler.

Figure 6.7 Frangible bulbs break when they reach a certain temperature, permitting water to flow from the sprinkler. The color of the liquid reflects the temperature rating.

sprinkler can be changed. Operating temperatures vary from 135°F to 575°F (57°C to 302°C).

The selection of the operating temperature of a sprinkler is determined by the maximum air temperature expected at the level of the sprinkler under normal conditions. For example, under conditions of ordinary room temperatures, a sprinkler with a temperature rating of 135 to 170°F (57°C to 77°C) is most frequently used. If the typical ambient temperature exceeds 100°F (38°C), such as in an attic or near a heater, a sprinkler with a higher temperature

rating is used. It is necessary to provide a margin between the normal room temperature and the operating temperature of a sprinkler because the solder will begin to yield as surrounding temperatures approach its melting point. It is also possible to vary the operating temperature of sprinklers using other types of heat-sensitive elements.

Temperature ratings that are too low for a given location may result in malfunction or accidental activation. Temperature ratings that are too high will delay sprinkler operation, enhancing fire growth. Some insurance carriers recommend the use of high-temperature sprinklers in industrial and warehouse occupancies as a way to limit the total number of sprinklers likely to activate during a fire. Insurance carriers recommend this practice because it is more effective in extinguishing or controlling the fire. It also reduces water damage, particularly by lessening the possibility of accidental activation due to a high ambient temperature.

Sprinkler frames or glass bulbs are color-coded so that their temperature ratings can be distinguished quickly **(Figure 6.8)**. The temperature rating is also stamped on the link in fusible link-type sprinklers. On other types of sprinklers, the temperature rating is stamped on some other part of the sprinkler **(Figures 6.9 a and b, p. 170)**. The sprinkler temperature ratings and color codes are listed in **Table 6.1, p. 170**.

Sprinkler Response Time

Because all sprinklers, with the exception of open deluge sprinklers, are thermal devices, there is some delay between the ignition of a fire and the

Figure 6.8 Temperature ratings can be quickly distinguished on color-coded sprinklers.

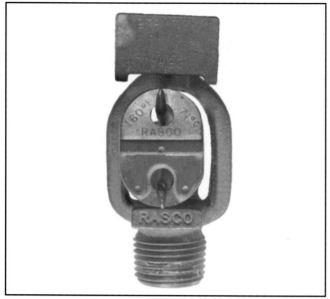

Figures 6.9 a and b Temperature ratings can be stamped or cast on some sprinklers.

operation of the sprinkler. This delay is a function of several variables, including the design of the sprinkler.

The activation time of the sprinkler depends on the surface area, mass, and thermal characteristics of the heat-sensitive element. Although the gas temperature surrounding a sprinkler may be 165°F (74°C), a heat-sensitive element rated at 165°F (74°C) will not reach this temperature for some time depending on its surface area, mass, and thermal characteristics. In addition, the temperature and velocity of a gas jet as it travels across the ceiling will affect the activation time because heat must be transferred from the hot gas jet to the operating element of the sprinkler. The faster and hotter the gas jet, the faster the sprinkler will operate. The relative speed of operation can be evaluated by a number known as the *Response Time Index (RTI)*. The lower a sprinkler's RTI, the faster it responds.

To speed the operation of sprinklers, engineers have designed a type of sprinkler known as an early-suppression fast-response (ESFR) sprinkler or a quick-response sprinkler. These faster-operating sprinklers can be compared to ordinary sprinklers through the use of the RTI. Standard-response sprinklers have RTIs that vary from 200 to 700 $(ft/sec)^{1/2}$ or $\{100 (m/sec)^{1/2}$ to $400 (m/sec)^{1/2}\}$. By improving its design, the RTI of the quick-response

Table 6.1
Sprinkler Temperature Ratings, Classifications, and Color Codings

Max. Ceiling Temp.		Temperature Rating		Temperature Classification	Color Code	Glass Bulb Colors
°F	°C	°F	°C			
100	38	135 - 170	57 - 77	Ordinary	Uncolored or black	Orange or red
150	66	175 - 225	79 - 107	Intermediate	White	Yellow or green
225	107	250 - 300	121 - 149	High	Blue	Blue
300	149	325 - 375	163 - 191	Extra high	Red	Purple
375	191	400 - 475	204 - 246	Very extra high	Green	Black
475	246	500 - 575	260 - 302	Ultra high	Orange	Black
625	329	650	343	Ultra high	Orange	Black

Reprinted with permission from NFPA 13, *Standard for the Installation of Sprinkler Systems*. Copyright© 2002 National Fire Protection Association, Quincy, MA 02269. This reprinted material is not the complete and official position of the National Fire Protection Association on the referenced subject, which is represented only by the standard in its entirety.

sprinkler can be reduced to about 50 (ft/sec)$^{1/2}$ or 28 (m/sec)$^{1/2}$, which results in operation that is approximately 5 to 10 times as fast. NFPA 13 defines fast-response sprinklers as having an RTI of 90 (ft/sec)½ {50 (m/sec)½} or less and standard response sprinklers as having an RTI of 144 (ft/sec)½ {80 (m/sec)½} or more.

Deflector Component

Another primary component of a sprinkler is the deflector, which is attached to the sprinkler frame. The deflector forms the discharge pattern of the water. Discharging water is directed against the deflector to convert it into a spray pattern.

Sprinklers produced before 1955 (now known as old-style sprinklers) were designed with deflectors that discharged a portion of the water upward toward the ceiling in order to protect structural elements. Unfortunately, this design did not produce a good downward distribution of water. Modern standard sprinklers produce a more uniform discharge pattern that is directed downward. By directing all of the water downward, the fire is controlled more effectively, resulting in reduced ceiling temperatures and better protection for structural elements. Because of their difference in discharge patterns, old-style sprinklers cannot be used to replace modern sprinklers. Modern sprinklers may be substituted for old-style sprinklers in an existing system if it becomes necessary to change or upgrade a sprinkler system.

In addition to being modified for specific functions, sprinklers can also be modified for different environments. For example, if a sprinkler is to be placed in a corrosive atmosphere such as a plating room, it can be coated with wax to protect it. If mechanical damage is possible, the sprinkler can be fitted with a protective cage (**Figures 6.10 a and b**). In areas where appearance is important, a variety of recessed, flush, and concealed sprinklers are available that may have a finish matching the color and texture of the ceiling. Such finishes must be applied by the sprinkler manufacturer, NOT the contractor or occupant.

The deflector configuration is fundamental to the effectiveness of the sprinkler. There are four basic types of sprinkler design: upright, pendant, sidewall, and water spray nozzles.

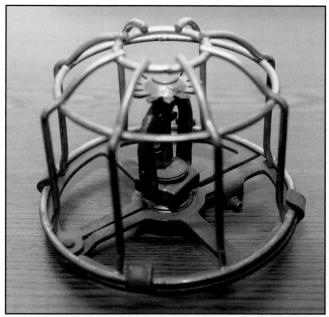

Figures 6.10 a and b A protective cage can help avoid damage to a sprinkler. Flushed or concealed sprinklers are used where appearance is a concern.

Upright

Upright sprinklers are designed to deflect the spray of water downward in a hemispherical pattern (**Figure 6.11, p. 172**). Upright sprinklers cannot be inverted for use in the hanging or pendant position; because the spray would be deflected toward the ceiling.

Pendant

Pendant sprinklers are used where it is impractical or unsightly to use sprinklers in an upright position, such as below a suspended ceiling (**Figure 6.12, p. 172**). The deflector on this type of sprinkler breaks the pattern of water into a circular pattern

Figure 6.11 Upright sprinklers will deflect the flow of water downward in a hemispherical pattern.

Figure 6.12 Pendant sprinklers also direct the flow of water downward.

of small water droplets and directs the water downward. Pendant sprinklers may be installed under wharves where it is desirable to direct the water upward. This practice is permitted by NFPA 13.

Sidewall
It is frequently desirable to install sprinklers on the wall at the side of a room or space for reasons of economy or appearance. By modifying the deflector, a sprinkler can be made to discharge most of its water to one side. These sidewall sprinklers are useful in such areas as corridors, offices, hotel rooms, and residential occupancies (**Figure 6.13**). For a variety of reasons, extended-coverage sidewall sprinklers have become more popular in certain occupancies. One reason their popularity has increased is that an extended-coverage sidewall sprinkler can be positioned at one end of a rectangular room, thereby reducing piping costs.

Water Spray Nozzle
Water spray nozzles discharge water in a specific pattern (typically conical) to protect a three-dimensional hazard such as an electric transformer or flammable liquid or gas tank. These sprinklers may or may not have a deflector and must be installed in a specific orientation to properly wet the surface of the hazard.

Figure 6.13 When sprinklers must be installed on a wall, sidewall sprinklers direct the flow of water to one side.

Early-Suppression Fast-Response Sprinklers
Fire sprinklers generally operate with an objective of fire control. Fire control is desirable but not as important as fire suppression. Early-Suppression Fast-Response (ESFR) sprinklers were designed to operate under the concept of fire suppression as the most desired result. ESFR sprinklers are a newer type of fast-response fire sprinkler. At press time, the generally accepted description of a true ESFR sprinkler is that four factors must be built into its design:

- Thermal sensitivity or RTI
- Actual delivered density (ADD)
- Deflector design
- Discharge orifice

Sprinklers that do not have all of these properties cannot be relied upon to achieve early fire suppression. ESFR sprinklers can be identified quickly because they are larger than conventional sprinklers.

Thermal Sensitivity

The response time for a sprinkler is determined by its sensitivity to heat, its operating temperature, and its distance from the fire, which affects the hot gas velocity and temperature. Activation of the sprinkler can be somewhat delayed if its sensitivity to heat (responsiveness) is slow. The fire gases can become higher than the temperature rating of the sprinkler, resulting in a larger fire plume and more difficulty in achieving early fire suppression. An ESFR sprinkler is designed to have a low RTI, thus ensuring fast activation. With quick activation, a fire is suppressed while it is smaller. The smaller the fire, the less demand for water.

Actual Delivered Density/Required Delivered Density

Actual delivered density is a measure of the amount of water discharged that actually reaches the fuel surface. ADD is determined by the following factors:

- Heat Release Rate of the fire, which affects the fire plume velocity and temperature
- Water drop size and momentum
- Distance of water drop travel to fire

Required delivered density (RDD) is the amount of water needed at the fuel surface to suppress a given fire. The RDD value is variable and depends on the size of the fire at the time of the operation of the sprinkler.

ESFR sprinklers are based on the theory that the earlier water is applied to a growing fire, the lower the *required* delivered density and the greater the *actual* delivered density will be. ESFR sprinklers are normally used in specific high-challenge fire hazards. For more details and requirements of

ESFR sprinklers, see NFPA 13, *Standard for the Installation of Sprinkler Systems*.

Deflector Design and Discharge Orifice

The design of the deflector, along with available pressure, affects the distribution pattern, the size, and the initial downward velocity of the water droplets. The discharge orifice is designed to discharge a large quantity of water. This also affects the size and velocity of the water droplets. It is the larger and faster droplets that penetrate the fire plume and reach the seat of the fire, thus increasing the actual delivered density. It is because of this key difference in the design of the deflector and the orifice that an ESFR sprinkler can discharge between 100 and 180 gpm (400 L/min and 720 L/min). In comparison, a standard ½-inch or $^{17}/_{32}$-inch sprinkler will discharge between 15 and 50 gpm (60 L/min and 200 L/min).

Sprinkler System Piping

The types of piping used in sprinkler systems are specified in the various NFPA sprinkler system standards. In general, there are four basic types of pipe: ferrous metal (steel, [sometimes called black steel]), copper tubing, galvanized, and plastic. Ferrous metal piping is the most common type in use (**Figure 6.14, p. 174**). The wall thickness of pipe is normally referred to as the pipe's *schedule*. Sprinkler piping is schedule 10, schedule 40, or special listed pipe. Schedule 10 may be joined by welding or with rolled-grooved fittings. Schedule 40 may be joined by welding, with rolled-grooved, cut grooved, or with threaded fittings. Steel piping has a long life expectancy but will eventually corrode and is very heavy.

Copper piping has been approved for sprinkler systems since the early 1960s. It is joined together by soldering or brazing. Copper piping is highly resistant to corrosion, has low friction loss, is lighter in weight than steel pipe, and is neat in appearance where exposed. It is rather expensive, however, and its use has declined with the increase of less-expensive plastic pipe.

Plastic pipe is the least expensive type of sprinkler system piping, the lightest in weight, and the easiest to install. Plastic pipe is joined with plastic cement. The main drawback of plastic pipe is that

Figure 6.14 Ferrous metal piping is the most common type of piping used for sprinkler systems.

Figure 6.15 As with any other type of sprinkler hardware, inspection personnel must be familiar enough with flexible sprinkler connections to determine that they have been installed correctly. *Courtesy of FlexHead Industries.*

it may not be installed in areas where the ambient temperature exceeds 120° F (49°C) for polybutylene (PB) and cross-linked polyethylene pipes, or 150°F (65°C) for CPVC pipes. It also must be enclosed behind a fire-resistant covering.

Newer flexible fire sprinkler connections are used to connect the sprinkler to the branch line using corrugated stainless steel tubing with a braid sheath (**Figure 6.15**). These connections eliminate the need to install hard-pipe armovers from the branch line to the sprinkler location. The benefits of these connections include quicker installation, ease of retrofits, and cost-effective code compliance.

NOTE: Polybutylene piping has not been widely used since the late 1980s in favor of the more accepted cross-linked polyethylene. Its use is also illegal in a number of states. For more information on placement and protection requirements for piping, see NFPA 13D and NFPA 13R.

The actual specifications of piping fall within the responsibilities of the plans reviewer or the FPE (fire protection engineer). The inspector needs to be able to recognize changes or modifications to the system.

Whether the sprinkler system has metal or plastic piping, the various types and sizes of pipes in a sprinkler system are given specific functional names based on the role each serves within the system. These names include the following:

- **Water supply main.** The *water supply main* is the piping that connects the sprinkler system to the main water supply (**Figure 6.16**). The water supply main is the underground municipal main that is used to supply other systems such as fire hydrants and fire pumps. The water supply main may also be the private fire main that connects the municipal main to the sprinkler system riser.

- **System riser.** The vertical piping that extends upward from the water supply to feed the cross or feed mains. The system riser will have a control valve unless an outside control valve such as a post indicator valve is used. The riser will have a drain, pressure gauge, and a waterflow alarm device attached to it.

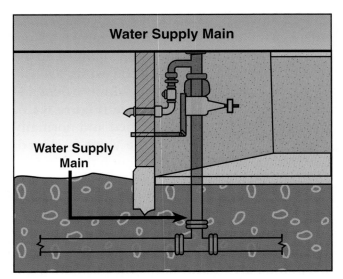

Figure 6.16 The water supply main is the underground municipal main that is used to supply other systems such as fire hydrants and fire pumps.

- **Sprig-up.** A pipe that rises vertically and supplies a single sprinkler.
- **Riser.** Any other vertical supply piping in the system is simply called a *riser* (**Figure 6.17**).
- Feed main. Pipes that supply water to each of the cross mains.
- Cross main. *Cross mains*, in turn, feed the branch lines.
- Branch line. Those pipes that contain the individual sprinkler devices and include grid lines on gridded systems. In ordinary-hazard occupancies, branch lines are usually limited to eight sprinklers. This amount may be increased to ten sprinklers if the pipe schedule is adjusted. Adjustment can be made by supplying the second sprinkler from the end with a 1¼-inch (32 mm) pipe and feeding the branch line with a 2½-inch (65 mm) pipe rather than a 2-inch (50 mm) pipe.

Valves

Every sprinkler system is equipped with various water control valves and operating (test and drain) valves. The following sections highlight the more common types of valves.

Control Valves

The main control valve is used to shut off the water supply to the system when it is necessary to replace sprinklers or to perform other maintenance. These

Figure 6.17 Vertical piping in the sprinkler system is called a riser.

valves are located between sources of water supply and the sprinkler system. After maintenance has been performed, the control valves must always be returned to the open position. Numerous major losses have occurred because the water supply to the sprinkler system was shut off when a fire broke out. To ensure that valves are returned to their normal fully open position, a *main drain test* is required by NFPA 25 when control valves are operated.

To help ensure that they are not inadvertently left closed, most valves that control water to sprinkler systems are the indicating type. With an indicating valve, the position of the valve – open or closed – can be determined at a glance. However, not all water control valves are indicating-type valves; underground valves in water distribution systems are commonly nonindicating valves.

There are three types of control valves: butterfly valves, gate valves, and ball valves. The most common type of indicating valve used in sprinkler systems today is the butterfly valve. Older systems typically used a gate valve called an outside stem & yoke or OS&Y valve. This valve is a gate valve that has a yoke on the outside with a threaded stem that controls the opening and closing of the gate by turning a handwheel **(Figure 6.18)**. The threaded portion of the stem is outside the yoke when the valve is open and inside the yoke when the valve is closed. The position of an OS&Y valve is easily seen from a distance as opposed to other types of indicating valves.

Other types of indicating gate valves are the post indicator valve (PIV) and the wall post indicator valve (WPIV). The PIV is used to control underground sprinkler valves and consists of a hollow metal post attached to the valve housing. The valve stem is inside this post. Mounted on the stem is a movable target with the words "OPEN" or "SHUT" visible through a window depending on the position of the valve **(Figure 6.19)**. The operating handle is fastened and normally locked to the post. When the valve is closed, the word "SHUT" appears through the window. The wall post indicator valve is similar to a PIV. However, the WPIV extends through the building wall with the target and valve operating wheel on the outside of the building **(Figure 6.20)**. The indicating butterfly valve has a paddle indicator or a pointer arrow that shows the position of the valve **(Figure 6.21)**.

Figure 6.18 On an OS&Y valve, the threaded portion of the stem is outside the yoke when the valve is open and inside the yoke when the valve is closed.

Figure 6.19 This post indicator valve (PIV) is in the open position.

Figure 6.20 Wall post indicator valves (WPIV).

Figure 6.21 The butterfly valve has an indicator that shows the position of the valve.

The actual valve mechanism of sprinkler system control valves may be of either the gate or the butterfly type. A gate valve mechanism consists of a close-tolerance gate that slides across the waterway (**Figure 6.22**). In a butterfly valve, a disc rotates 90 degrees *inside* the waterway (**Figure 6.23**). Butterfly valves are operated by a worm gear, which is turned by a handle or a handwheel. The position of a butterfly valve is indicated by a pointer that points either to the word "OPEN" or "CLOSED" on the valve body or by a cross-view of the valve showing its open or closed position.

A ball valve may be used as a control valve as long as it cannot be closed in less than 5 seconds. Closing any valve in less than 5 seconds is always ill-advised because it may cause water hammer,

resulting in damage to system components. Ball valves that are listed for use in a fire sprinkler system as a control valve have a handwheel. These valves are typically found as floor control valves. The ball valve has an indicating device on the body of the valve indicating whether or not the valve is open.

In order to further ensure that sprinkler valves are kept open, the valves are either chained or locked in the open position or are electrically supervised. A valve that is electrically supervised has a switch attached to it. Movement of the valve beyond two turns of the handwheel or one-fifth the total travel of the valve causes an electrical circuit to open and transmit a signal to a watch service or manned guard post.

Operating Valves

In addition to the control valves, sprinkler systems employ various operating valves such as check valves, automatic drain valves, globe valves, and stop or cock valves.

Check Valves

Check valves are used to limit the flow of water to one direction. They are placed in water sources to prevent recirculation or backflow of water from the sprinkler system into the municipal water supply

Figure 6.22 When the system is closed, the gate valve will slide down to close the waterway.

Figure 6.23 The butterfly valve rotates inside the waterway.

system. Check valves usually have an arrow cast into the body to indicate direction of flow (**Figure 6.24**). If no arrow or indication of flow direction can be seen on the exterior of the valve, the valve clapper pivot should be on the end toward the source of supply (**Figure 6.25**). If the arrow is pointing toward the water source or the pivot is on the sprinkler riser side, the check valve has been improperly installed.

Figure 6.24 The direction of water flow can be seen by the arrow cast on the exterior of the check valve.

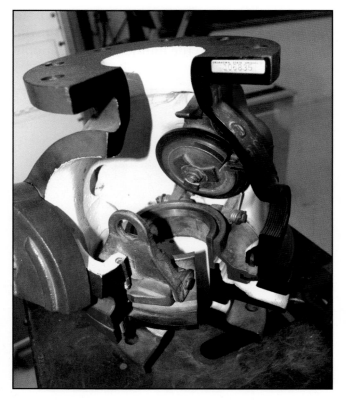

Figure 6.25 The valve clapper pivot should be on the end toward the source of supply.

Although check valves are simple devices, occasionally they may fail to seat (reset) properly. Likely causes of an improper seat are a bad rubber seal, obstructions, or a sticking clapper. If water leaks from the fire department connection, suspect an improperly seated check valve and advise building authorities that repairs are needed.

Backflow Prevention Devices

Many municipal governments, health departments, private water purveyors, and water quality assurance agencies require the use of backflow prevention devices between the public water main and automatic fire sprinkler system. Any fire department connection (FDC) that is used to boost the pressure of the sprinkler system from a fire department engine must be situated between the backflow prevention device and the sprinkler riser (**Figure 6.26**). A listed backflow prevention device can be considered a substitute for the sprinkler check valve as required on all automatic fire sprinkler systems. The sprinkler system designer must remember to include the pressure loss charge created by the backflow prevention device during the design process. This loss is usually between 10 and 20 psi (70 kPa and 140 kPa) and is dependent upon the size and capacity of the device.

Automatic Drain Valves

Main drain and auxiliary valves drain piping when pressure is relieved in the pipe. Their most common application is to drain water from siamese connections after use in dry-pipe, preaction, and deluge valves. This prevents the portion of the pipe that extends through the wall from freezing in cold weather.

Globe Valves

Globe valves are small handwheel-type valves that are typically designed to be turned counterclockwise to open and clockwise to close. Globe valves are used primarily on drains and test valves. The *inspector's test valve* is usually a globe valve and may sometimes be found at a remote location within a sprinkler system (**Figure 6.27**). The inspector's test valve will be opened by an inspector to cause water to flow through the system. This test is performed to ensure that the audible alarm

Figure 6.26 Note that the fire department connection (FDC) is placed between the backflow prevention device and any sprinkler system.

Figure 6.27 A typical inspector's test valve.

activates in a timely fashion and is indeed audible. In some modern sprinkler system designs, the inspector's test valve is located near the riser for convenience.

Ball-Drip Valves

Ball-drip valves, sometimes known as stop or cock valves, are also used for drains and alarm testing. Ball-type valves are most commonly opened or closed with a quarter turn of the valve.

Fire Department Connections (FDCs)

In addition to the primary sources of water supply, sprinkler systems are provided with one or more fire department connections (sometimes called siamese connections or simply the FDC). These fire department connections permit the fire department pumper to pump into the system, thereby boosting the pressure and the volume of water in the system **(Figure 6.28, p. 180)**. Fire department connections are especially important to the protection of a building in which the primary water supply is weak, the supply may be overtaxed when a large number of sprinklers opens, or where the system is old and heavy pipe corrosion is suspected.

The most common type of FDC consists of a 4-inch (100 mm) pipe equipped with a siamese fitting on the outside of the building. The siamese fitting has two 2½-inch (65 mm) female inlets to which fire hoses are connected. Large systems may have three or more inlets **(Figure 6.29)**. Some newer systems are equipped with inlets for large diameter hose (4- or 5-inch [100 mm or 125 mm]).

Figure 6.28 Fire department connections (FDC) enable a fire department pumper to boost the pressure and/or supply of a sprinkler system. *Courtesy of Ted Boothroyd.*

Figure 6.29 Large sprinkler systems may have a number of FDC inlets. *Courtesy of Ted Boothroyd.*

Where there is a riser control valve, a check valve is typically provided inside the building on the 4-inch (100 mm) pipe to keep system water out of the portion of the pipe that extends outside the building. This prevents damage from freezing temperatures and keeps the system pressure off the clappers in the hose inlets. An automatic drain valve (ball-drip valve) is also installed in the FDC to prevent problems associated with freezing temperatures. The check valve may also be in the form of a backflow preventer installed as part of the fire service underground. The FDC may be installed as a free-standing FDC or on the downstream side of the backflow preventer.

On a single-riser sprinkler system, the fire department connection is attached on the sprinkler system side of the main supply valve. On multiple-riser systems, the FDC enters the supply piping between the main supply valve and the individual riser valves. This arrangement allows water to be supplied to the system even when one riser valve has been closed deliberately or inadvertently; water will still be pumped into the remaining risers. Water cannot reach the affected area, however, if floor or sectional valves are closed. **Figure 6.30** shows show the proper arrangement of the FDC relative to the sprinkler system. The FDC is also located on the building/sprinkler side of the backflow prevention valve.

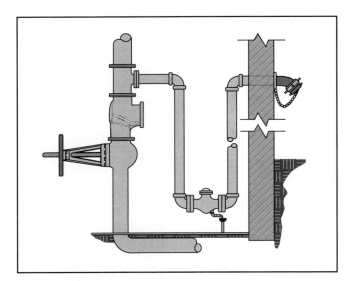

Figure 6.30 The piping arrangement for a fire department connection on a wet-pipe sprinkler system.

Types of Sprinkler Systems

The more common types of sprinkler systems in use today include the following:

- Wet-pipe system
- Dry-pipe system
- Deluge system
- Preaction system

Wet-Pipe Sprinkler System

The wet-pipe system is the oldest, simplest, and most reliable type of sprinkler system **(Figure 6.31)**. This type of system contains water under pressure in the piping at all times so that the opening of a sprinkler immediately discharges water onto a fire and actuates an alarm. Depending on the size of the systems and the occupancies in which they are built, wet-pipe systems fall under the requirements of NFPA 13, 13D, or 13R. Standard wet-pipe systems may be installed in any location where system components will not be subject to freezing. Wet-pipe systems of limited size may be installed in areas subject to freezing if they are freeze-protected.

NOTE: Antifreeze systems are addressed later in this section.

Wet-pipe systems are usually successful in controlling fires with only a minimum of sprinklers opening. Records kept by the NFPA indicate that the vast majority of fires are controlled with less than 10 sprinklers activated. These statistics highlight the reliability and effectiveness of sprinkler systems. This type of system can be equipped with an alarm check valve and a water gong or pressure switch. Alternatively, a waterflow indicator may be installed in the riser. These devices initiate an alarm when water begins to flow in the system.

Alarm Check Valve

Some wet-pipe systems include an alarm check valve on the riser. An alarm check valve has been specially designed to incorporate an alarm function **(Figure 6.32)**. When one or more sprinklers open, the flow of water lifts the main clapper from its seat and uncovers a pipe that directs some of the water to a pressure switch and/or a water motor gong, thus activating the alarm.

Figure 6.31 Wet-pipe systems can be installed in areas not subject to freezing.

Figure 6.32 Sprinkler systems can be designed to send an alarm as they begin to discharge water.

Because water pressure can vary, causing "surges" that may inadvertently trigger an alarm, a retard chamber is provided on systems with alarm check valves to prevent false alarms. The retard chamber is a small tank made of cast iron or pressed steel that is installed in the alarm pipe to absorb a small amount of water flow during pressure fluctuations, thus preventing unnecessary alarm operation. The amount of water that the retard chamber absorbs usually prevents surges from triggering alarms, but it does not significantly reduce the alarm response time during an actual activation of the system. The retard chamber is provided with a small drain so that water from surges does not accumulate (**Figures 6.33 a and b**).

The alarm check valve is also equipped with a test valve. This small valve diverts water from the water supply into the alarm pipe and retard chamber, enabling alarms to be tested without tripping the alarm check valve.

Waterflow Switch

The waterflow switch consists of a vane or paddle that protrudes through the riser into the waterway (**Figure 6.34**). The vane is made of thin, flexible plastic or metal and is connected to an alarm

Figure 6.33 a The amount of water that collects in this retard chamber is designed to prevent unnecessary alarms. *Courtesy of Ted Boothroyd.*

Figure 6.33 b The flow of water through a wet-pipe system equipped with a retard chamber.

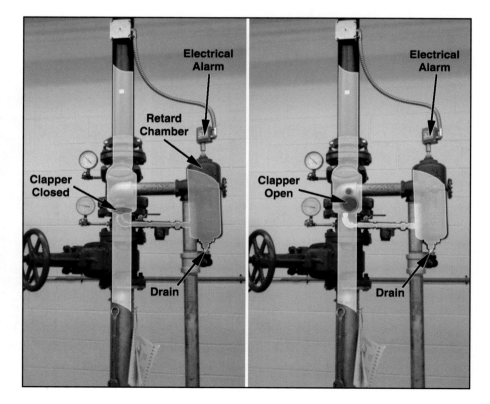

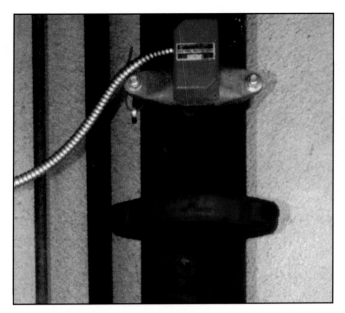

Figure 6.34 Waterflow switches can also be provided with a time-delay function to prevent unnecessary alarms. *Courtesy of Ted Boothroyd.*

Figure 6.35 a A waterflow switch used to localize the origin of an alarm.

Figure 6.35 b Valves installed to isolate sections of a sprinklered building.

switch located on the outside of the riser. Movement of the vane, which is caused by flowing water, operates the switch that initiates the alarm. The vane must be sufficiently thin and pliable so that if many sprinklers operate, the water flow will flatten the vane against the inside of the riser, resulting in a clear waterway.

Waterflow switches can be provided with a time-delay feature to prevent false alarms — typically 0 to 30 seconds from activation — just as the retard chamber does with an alarm check valve. Waterflow switches cannot be used in dry-pipe systems, deluge systems, or preaction systems because the sudden rush of water that occurs when these systems operate could damage the vane. Waterflow switches are compact and frequently used where only an electrical output from an alarm is desired.

Vane-type waterflow switches are used as sectional or floor indicators to localize the origin of an alarm. This arrangement is frequently found in high-rise buildings. In addition, sectional valves can be installed to isolate sections of the sprinklered building without interrupting protection to the entire building. Examples of such arrangements are shown in **Figures 6.35 a and b**.

Local and Fire Department Notification Systems

The purpose of any sprinkler alarm is to notify someone that the sprinkler system has activated as well as to notify the fire department. Prompt notification of the fire department is important because the sprinklers may not completely extinguish the fire. Furthermore, systems occasionally leak because of frozen pipes or mechanical damage, so prompt notification will minimize water damage.

Once the system has operated and the waterflow alarm has sounded, the actual alarm signal can take several forms. An electric pressure switch or waterflow switch can be used to transmit the alarm to several locations. The signal can be transmitted to the local fire department directly, to a supervising station, or simply to local bells installed on the protected property. The types of supervising stations include central station, proprietary station, remote station, and auxiliary station.

Another type of local alarm is the water motor gong (**Figure 6.36**). The water motor gong consists of a local alarm bell that is controlled by the action of water that flows through a small waterwheel after it leaves the retard chamber. The waterwheel operates a gong located on the outside of the building. Good practice is to locate the gong over the fire department connection to assist responding firefighters in locating the connection. Water motor gongs are subject to freezing and require draining of the alarm water line after activation.

Antifreeze Systems

When a wet-pipe sprinkler system must protect such small, unheated areas as outside loading docks, an antifreeze system may be used (**Figure 6.37**). In an antifreeze system, the ordinary water in the sprinkler piping is replaced with a nonfreezing, nonflammable liquid. Increased concern for environmental health has resulted in public health regulations restricting the type of chemicals that can be used in any piping system connected to the public water supply. However, chemically pure or United States pharmacopoeia-grade glycerin or propylene glycol is usually acceptable. In systems not connected to a public water supply, other liquids such as diethylene glycol, ethylene glycol, propylene glycol, or calcium chloride can be used. The cost of the antifreeze liquids usually restricts their use to systems of less than 40 gallons (160 L). Alternatives to antifreeze systems are preaction systems, dry sprinklers, and dry-pipe systems.

Because the specific gravity of antifreeze solutions is greater than water, an antifreeze system is constructed in the shape of a U-loop (**Figure 6.38**). The U-loop should have a control valve, a fill cup,

Figure 6.36 A hydraulic sprinkler waterflow alarm.

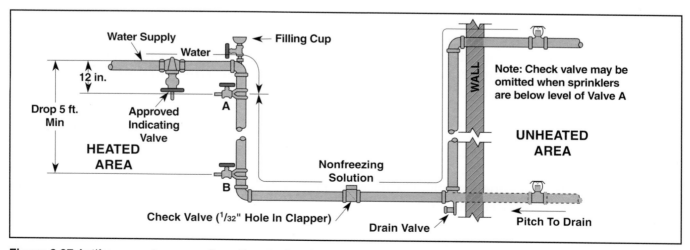

Figure 6.37 Antifreeze systems are alternatives to dry-pipe systems. Reprinted with permission from NFPA 13-2002, *Standard for the Installation of Sprinkler Systems, Copyright© 2002 National Fire Protection Association, Quincy, MA 02269. This reprinted material is not the complete and official position of the National Fire Protection Association on the referenced subject, which is represented only by the standard in its entirety.*

Figure 6.38 This piping arrangement prevents the antifreeze solution from leaking back into the rest of the system.

Figure 6.39 A typical dry-pipe valve and riser arrangement.

two solution test valves, a drain valve, and a check valve. This arrangement prevents dilution of the antifreeze solution and keeps the antifreeze out of the main portion of the system.

The condition of the antifreeze solution should be checked each year before the beginning of freezing weather. This check can be performed by emptying the system and checking the solution's specific gravity. A danger with an antifreeze system is the possible use of flammable or toxic liquids by unskilled personnel.

Dry-Pipe Sprinkler System

A dry-pipe sprinkler system is one in which air or nitrogen under pressure replaces the water in the pipes exposed to freezing temperatures. When buildings lack sufficient heat to keep the water in the sprinkler pipes from freezing, a dry-pipe sprinkler system must be used **(Figure 6.39)**. The dry-pipe valve must be located in a heated portion of the structure, which must be maintained at a

minimum of 40°F (5°C) according to NFPA 13 and NFPA 25. If the valve is subject to freezing conditions, it may not activate properly when required. In some cases, it may be necessary to build a small, heated enclosure specifically for the dry-pipe valve riser and its associated equipment.

A device known as a dry-pipe valve uses the force created by the air pressure in the system to hold the clapper in the valve closed and to keep water from entering the system until a sprinkler operates. The reduction in air pressure due to air escaping from the open sprinkler permits the dry-pipe valve to open and admit water to the system **(Figure 6.40, p. 186)**. Dry-pipe systems can be equipped with either electric or hydraulic (water) alarm-signaling equipment.

The air necessary to service a dry system may be obtained from two sources. One source is to have an air compressor/tank arrangement dedicated for exclusive use in the system. The second alternative is to design the sprinkler system into a monitored facility-wide compressed-air system.

Dry-Pipe Valves

Dry-pipe valves are designed so that a small amount of air pressure above the dry-pipe valve holds back a much greater water pressure on the

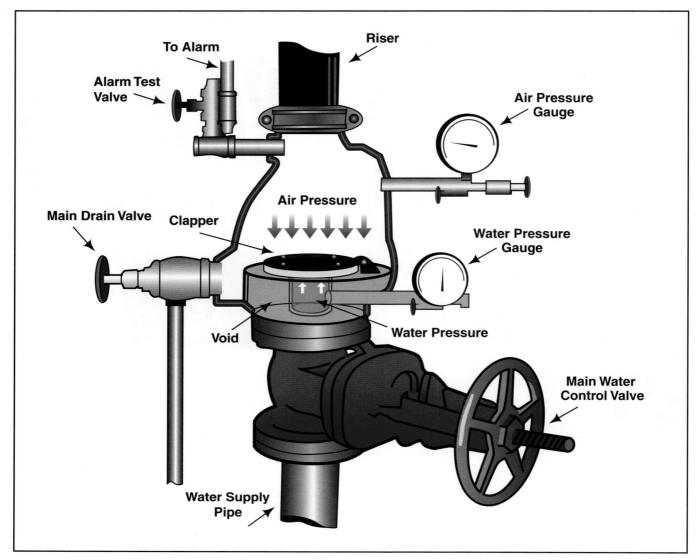

Figure 6.40 An internal view of a differential dry-pipe valve.

Labels in figure:
To Alarm
Riser
Alarm Test Valve
Air Pressure Gauge
Main Drain Valve
Clapper
Air Pressure
Water Pressure Gauge
Void
Water Pressure
Main Water Control Valve
Water Supply Pipe

water side of a valve. This action is accomplished through the differential principle incorporated in the valve design. When the differential dry-pipe valve operates and the clapper swings open, the valve is held in place by a latch to prevent its reseating. Releasing the latch requires that the valve body be opened; thus, ordinary dry-pipe valves must be manually reset after they have operated.

The air-water differential serves two purposes. First, a comparatively small amount of air pressure is required to hold back the water. Second, the less air pressure in the system, the more quickly the water can expel the air when the dry-pipe valve operates. Therefore, the time needed for water to be discharged is reduced.

The air pressure in the system must be maintained within prescribed limits; typically, it must not be more than 20 psi (140 kPa) above the pressure at which the dry-pipe valve will trip. Higher air pressures are difficult to maintain, and operation of the valve is delayed considerably if excessively high pressure is used. In systems where pressure from the primary water supply is low and a fire pump is installed, sufficient air pressure must be maintained to keep the dry-pipe valve from tripping accidentally. Accidental tripping of the valve can result from the high water pressure that is created when the fire pump operates.

Typically, a small amount of water is found above and on the air side of the dry-pipe valve clapper **(Figure 6.41)**. This water is necessary for priming

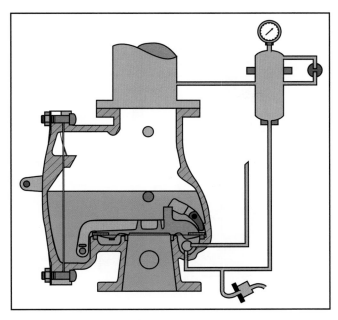

Figure 6.41 A small amount of priming water is found above the clapper.

purposes. If too much water enters the pipe above the valve, however, it can exert enough pressure against the air seat to hold the valve closed. This condition is called *water columning*. Automatic drip-valve drains are provided in the dry-pipe valve intermediate chamber to ensure that this condition does not occur. This drain will automatically close when the system is activated.

Although the ordinary dry-pipe valve is widely used, it has two shortcomings. First, it is slow to operate. Second, when the valve trips, the water, which is at a pressure significantly higher than the air pressure, rushes in with considerable velocity. This inrush of water can cause damage to the system components. The use of low differential dry-pipe valves will overcome these problems. Low differential dry-pipe valves use air pressure in the system piping at approximately 110 percent of the maximum static water pressure to hold back the water. These valves resemble alarm or check valves and have pressure differentials of 1.1 to 1 air-to-water pressure.

The following are several advantages of the low differential valve:

• Low velocity water entry

• Valve trip time 85 percent quicker than an ordinary valve

• Total flow time reduction of nearly 70 percent over an ordinary valve

• Easy to reset and can be used as either wet or dry valves

The disadvantages are that a heavy-duty air compressor is needed and an automatic pressure maintenance device is required.

Quick-Opening Devices

In a large dry-pipe system, several minutes can be lost while the air is being expelled through the open sprinklers before the dry valve operates. This delay provides time for fire growth, resulting in additional sprinklers opening and creating a greater demand. It is possible to speed the operation of the system by using a quick-opening device. There are two types of quick-opening devices: accelerators and exhausters. The accelerator unbalances the differential in the dry-pipe valve, which causes it to trip more quickly. The exhauster functions by quickly expelling air from the system.

NFPA 13 requires a quick-opening device on any dry-pipe valve that serves a piping system with a capacity of more than 500 gallons (2 000 L). Quick-opening devices are complicated mechanisms that require proper care and maintenance. NFPA 25 specifies that these devices should be tested four times a year by a qualified sprinkler inspector. Quick-opening devices are designed in such a way that if they fail to operate it will not prevent the dry-pipe valve from operating; it will just take longer.

In a system equipped with an accelerator, when a sprinkler is fused and the air pressure drops slightly (usually 2 pounds), a diaphragm in the accelerator becomes unbalanced (**Figures 6.42 a and b, p. 188**). This unbalanced condition causes a valve to open, which permits the air pressure in the system to enter the intermediate chamber of the dry-pipe valve. As soon as the air pressure is equalized on both sides of the clapper (normally 10 to 15 seconds), the valve is tripped open by the water pressure.

In a system equipped with an exhauster, the fusing of a sprinkler causes a diaphragm to open a large valve (**Figures 6.43 a and b, p. 188**). This action permits air pressure to escape quickly to

Dry-Pipe Accelerator

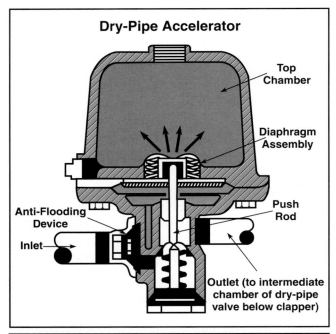

Top Chamber

Diaphragm Assembly

Anti-Flooding Device

Inlet

Push Rod

Outlet (to intermediate chamber of dry-pipe valve below clapper)

Dry-Pipe Exhauster

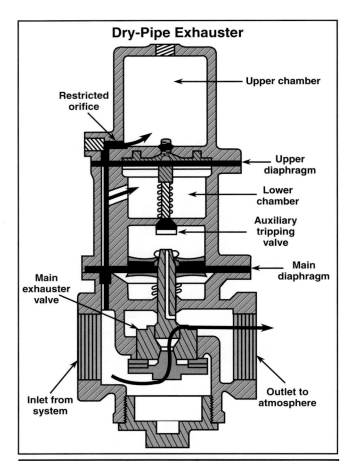

Restricted orifice

Upper chamber

Upper diaphragm

Lower chamber

Auxiliary tripping valve

Main diaphragm

Main exhauster valve

Inlet from system

Outlet to atmosphere

Figures 6.42 a and b An accelerator allows the dry-pipe valve to open within 10 to 15 seconds.

Figures 6.43 a and b The exhauster speeds the release of air from the system.

the outside and the dry-pipe valve to trip. After the system trips, the exhauster valve closes so that water is not lost from the system.

Operational Sequence for Dry-Pipe Systems

The basic operational sequence for a dry-pipe system is as follows:

1. Heat from a fire causes the sprinkler to activate.

2. Pressurized air contained in the piping begins to flow through the open sprinkler.

3. After a slight drop in air pressure, the quick-opening device (if present) activates to accelerate the removal of air from the piping (**Figure 6.44**).

4. Once the air pressure is reduced sufficiently, the dry-pipe valve trips open. The interior clapper is held in the open position by a latch.

5. Water enters the intermediate chamber of the dry-pipe valve. This automatically forces the automatic drip valve closed and begins the flow of water through alarm-signaling equipment.

6. Water flows through the entire piping system and is discharged through the open sprinkler.

Deluge Sprinkler System

A deluge system is similar to a dry-pipe system in that no water is contained in the piping before the activation of the deluge valve. However, it differs from a dry-pipe system in that all of the sprinklers are open, with no fusible links. This means that when the valve is tripped and water enters the system, the water will discharge through all the sprinklers simultaneously. The flow of water to the system is controlled by a deluge valve. Fire detection devices, which may be heat, smoke, or flame detectors, are installed in the same area as the sprinklers. The detection devices control the operation of the deluge valve through a tripping device and are required to be automatically supervised. Unlike wet- or-dry-pipe systems, the sprinklers do not function as the detection device in a deluge system.

A deluge system is designed to quickly supply a large volume of water to the protected area. Deluge systems are normally used to protect extra-hazard

Figure 6.44 An installed dry-pipe exhauster.

occupancies such as aircraft hangars, woodworking shops, cooling towers, ammunition storage, or certain manufacturing facilities. Deluge systems are sometimes used with foam to protect flammable liquid hazards. A system using some open and some closed sprinklers is a variation of the deluge system. Because deluge systems require a large volume of water, they are sometimes supplied by fire pumps.

Operation of the Deluge Valve

A deluge valve holds the water back until it is released. Just as there are several types of fire detection systems, there are also several methods of operating the deluge valve. One type of deluge valve that uses a weight to unlatch the valve is shown in **Figures 6.45 a and b, p. 190.** Deluge valves can be operated electrically, pneumatically, or hydrauli-

Figure 6.45 b When the system operates, a loss in pressure allows the valve to open and charge the system.

Figure 6.45 a Water pressure maintains the valve in a closed position.

cally. In addition, deluge valves have provision for manual operation. When a deluge system has more than 20 sprinklers, the detection devices must be electronically supervised.

Electrical operation. An electrically operated deluge valve is designed for use with fire detectors that transmit an electrical signal to the valve. The activation system includes an electrical tripping mechanism that releases the deluge valve clapper. Also included is a manual release and reset in case of power failure. Normal electrical service is a primary power source; however, batteries are usually provided as a secondary power source. A major advantage of the electrically operated deluge valve is its speed of operation. Most modern deluge systems are activated electrically.

Pneumatic operation. Pneumatically operated deluge valves are designed for use in a pneumatic detection system, usually one that has rate-of-rise detectors. This type of activation is actually a combination of pneumatic and mechanical activation. A pneumatic detector operates on an imbalance of pressure. Activation of the deluge valve occurs when a change in air pressure within the device, which is caused by an increase in temperature, displaces a diaphragm. The diaphragm is mechanically linked to a tripping mechanism that in turn

releases the deluge valve clapper. The pneumatic system does not require electrical power for activation; however, a manual release is required.

Hydraulic operation. There are several different types of hydraulically operated deluge valves **(Figure 6.46)**. Some of these valves may use a combination of a hydraulically operated valve and a pneumatic, hydraulic, or electrical detection system. Usually, no matter what combination of detection system is used, hydraulic pressure is required for activation by changing the deluge valve differential pressure.

Operational Sequence for a Deluge System
The following section highlights the sequence of operation for a deluge sprinkler system:

1. A product-of-combustion (heat, smoke, or flame) detector senses the presence of a fire condition, *or* an individual in the area discovers a fire in progress.

2. The fire detection system sends a signal to the deluge valve, causing the valve to open, *or* the individual who discovers the fire manually trips the deluge valve.

3. As water enters the deluge valve and the piping, a pressure switch is activated that transmits an

Figure 6.46 A typical deluge valve assembly.

Figure 6.47 A preaction valve is similar to a deluge valve.

alarm either locally or to a supervising station. A water motor gong (if present) is activated.

4. Water flows through all open sprinklers simultaneously.

Preaction Sprinkler System

A *preaction* system is used when it is especially important that water damage be minimized, even if the sprinkler pipes accidentally break (**Figure 6.47**). Preaction systems are frequently used to protect computer rooms, document storage areas, and freezers or cold-storage warehouses. The system employs a deluge-type valve, fire detection devices, and CLOSED sprinklers. Like deluge systems, the preaction system valve will not discharge water into the sprinkler piping until an indication is received from fire detection devices (other than the sprinklers) that a fire may exist. Once water is in the system, it may then be discharged through any sprinkler that has opened.

The piping in a preaction system is normally dry; therefore, it can be used in freezing environments such as cold-storage warehouses. Where the preaction system exceeds 20 sprinklers, NFPA 13 requires the piping to be supervised. Usually, air under a low pressure (the minimum required is 7 psi [49 kPa]) is maintained in the piping as a supervisory function. In the event of a leak or

break in the piping, the supervisory air pressure drops and transmits an alarm without admitting water to the system.

Preaction sprinkler systems fall into three categories:

- **Single interlock system**. A single interlock system admits water to sprinkler piping upon operation of detection devices.
- **Non-interlock system.** A non-interlocking system admits water to sprinkler piping upon operation of detection devices or automatic sprinklers.
- **Double interlock system.** A double interlock system admits water to sprinkler piping upon operation of both detection devices and automatic sprinklers.

Operational Sequence for Preaction Systems
The following gives the sequence of operation for a preaction sprinkler system:

1. A product-of-combustion (heat, smoke, or flame) detector senses the presence of a fire condition.

2. The fire detection system sends a signal to the preaction valve, causing the valve to open.

3. Sensors in the piping system detect the flow of water into the system and trigger the waterflow fire alarm.

4. When the level of heat at a sprinkler reaches the appropriate temperature, the sprinkler opens and water flows through the orifice.

Sprinkler Systems for Storage Occupancies

The storage practices that have evolved in modern warehouses make them a special challenge for both sprinkler system designers and firefighters **(Figure 6.48)**. Because of the enormous size of modern warehouses and the different methods of storing inventory, warehouses present severe fire control problems.

Fires in warehouses involve very heavy fire or fuel loads. Fires in these structures can develop rapidly and overtax inadequately designed sprinkler systems, resulting in extensive structural collapse and extremely high dollar losses. There may be a higher potential for damage in warehouses than in other portions of an industrial complex. It is also common to find retail stores with high-piled storage that look much like warehouses. A retail store of this type poses an even greater need for an adequate and swift response sprinkler system because of the potential for hundreds of people doing business inside the building. The higher loss potential can be attributed to several general conditions:

- Large undivided areas
- Large concentration of value in a single fire area
- Very high stockpiling with narrow aisles **(Figure 6.49)**.

Many stored items, such as machine parts, are only moderately combustible or are noncombustible. However, they are often packaged in cardboard containers and foam plastics. Other, more traditional packing materials include shredded paper, excelsior, and straw, which are also highly combustible. Therefore, much of the volume of

Figure 6.48 Fires in bulk paper storage can present a significant challenge to firefighters.

Figure 6.49 High-piled materials increase the loss potential in a warehouse.

a storage area can be made up of fast-burning material even when the basic products are non-combustible.

It is typical for modern warehouses to be constructed one story in height with very high ceilings. Rack storage up to 25 feet (8 m) is very common. However, automated warehouses using mechanical stacking equipment may have racks much higher than 25 feet (8 m). Such arrangements usually have very narrow aisles. The undivided area of a warehouse can vary from 100,000 square feet (9 290 m²) to 1,000,000 square feet (92 900 m²).

Sheltered voids that exist within high-piled palletized and rack storage promote fast fire spread out of the reach of conventional sprinkler systems. To protect these warehouses effectively, firefighters should have knowledge of the sprinkler systems and the storage methods associated with the warehouse.

With the exception of just a few occupancies, the standards of NFPA 13 apply to all occupancies in regard to the installation of sprinkler systems in storage occupancies. NFPA 230, *Standard for General Storage,* covers non-sprinkler requirements for storage occupancies including rubber tires, roll paper and a variety of chemicals.

Classification of Commodities

Identifying the commodities that are stored and the method of their storage are essential tasks for analyzing the amount of sprinkler protection needed for high-storage areas. The significant fire-related properties of a commodity include the following:

- Heat of combustion
- Rate of heat release
- Rate of flame spread

In addition, fire performance is affected by the storage configuration and material used for packaging. Packaged goods must be considered as a whole because that is the way they burn.

In determining protection requirements, stored commodities are divided into general hazard classifications based upon the behavior of typical items in each classification. A commodity consists of the product, the package (typically a cardboard carton or wood box), and any packaging aids such as plastic "peanuts" or Styrofoam™ blocks. The following are descriptions of commodity classifications:

- *Class I* — These products are generally noncombustible and stored on wooden pallets in ordinary packaging. Examples are machine parts, empty cans, metal cabinets, and appliances. A Class I commodity may also be packaged in corrugated cardboard or stretch-wrapped as a unit load.

- *Class II* — Also a noncombustible commodity but packaged in wooden crates or multilayered cardboard cartons.

- *Class III* — These commodities are made of combustible materials such as wood, paper, or certain plastics, irrespective of pallets or packaging.

- *Class IV* — A Class IV commodity meets one of the following criteria:
  — Constructed of Group B plastics.
  — Consists of free-flowing Group A plastics.
  — Contains an appreciable amount of Group A plastics. Remaining materials may be wood, metal, paper, natural or synthetic fibers, or Group B and C plastics.

Plastics present a special fire control problem because they produce more heat per unit of weight than ordinary combustibles. In recognition of the special nature of plastics, they are divided into three groups — A, B, and C — which are separate from the four commodity classifications previously described. Group A includes the fastest burning plastics, Group B is for moderate burning plastics, and Group C, the slowest burning plastics (**Table 6.2**).

Table 6.2 Types of Plastics		
Fastest Burning **Group A**	Moderate Burning **Group B**	Slowest Burning **Group C**
Acrylic	Cellulose Acetate	Melamine
Polycarbonate	Ethyl Cellulose	Phenolic
Polyethylene	Nylon	Polyvinyl Fluoride
Polystyrene	Silicone Rubber	Urea Formaldehyde

The classification (I, II, III, IV [plastics]) for storage is based upon the most severe hazard in the storage area. Miscellaneous types of storage, such as those found in automotive or hardware warehouses, have commodities in several classes. These are generally assigned a Class IV or plastic commodity classification because there is an appreciable amount of plastics being stored. It is often better to assign the next higher class than is originally contemplated because of the possibility of a higher-class commodity being introduced later. The difference in protection cost for the initial system design would be small compared to the cost of upgrading the sprinkler system at a later date.

Storage Methods

Higher-hazard commodities can sometimes be segregated to lessen their influence on the overall hazard class. For example, exposed foamed plastics or labeled flammable liquid drums can be placed in special rooms having greater fire protection and fire separation from the main storage area. Minor quantities of hazardous substances, such as cases of cigarette lighter fluid in a grocery warehouse or small propane cylinder hand torches in a hardware warehouse, tend to intrude without notice by warehouse personnel. It is not unusual to accept small quantities of such items in small containers without classifying the entire storage as flammable liquids or gases.

NOTE: Refer to **Appendix G, Storage Methods Used in Warehouses,** for more information about commodity storage.

Inspecting and Testing Sprinkler Systems

Sprinkler systems require periodic inspections and maintenance in order to perform properly during a fire situation. Plant managers, maintenance personnel, and fire prevention and inspection personnel should be able to inspect systems and identify problems **(Figure 6.50)**. Competent plant personnel or a contracted sprinkler company should perform routine maintenance. **Fire department personnel should not attempt to perform maintenance procedures if warranted by an inspection because this can result in possible liability if the**

Figure 6.50 Fire department personnel should be able to inspect sprinkler systems and point out areas that need attention.

system fails or a component breaks. Fire department personnel should, however, be able to point out problems in order to assist building owners in maintaining system readiness. Additionally, they should witness system tests for purposes of pre-incident familiarity and to verify system readiness.

Inspection personnel should take the following important steps before performing any inspection or witnessing any tests on a sprinkler system:

- Review the records of prior inspections and identify the make, model, type of equipment, and the area protected by the system.

- Wear appropriate clothing for dirty locations such as attics and basements. Protective clothing may be necessary for certain manufacturing areas **(Figure 6.51)**.

- Obtain permission from the plant management before performing any inspection. The inspection should *never* be performed without this approval.

- *Never* operate, adjust, manipulate, alter, or handle any sprinkler devices or equipment during situations other than emergencies or planned training sessions **(Figure 6.52)**.

- Notify the alarm-monitoring organization *before* testing has begun and *after* testing is completed if equipment is electronically supervised.

When testing is completed, the alarm-monitoring organization should confirm that the alarm equipment functions properly. If no alarms were received (waterflow, valve supervision, etc.), corrective maintenance service is required on the alarm system.

Figure 6.52 It is not the job of inspection personnel to manipulate or adjust any components of a sprinkler system. These functions should be carried out by maintenance personnel or building representatives.

Sprinklers

When changes occur in an occupancy or fire hazards within the occupancy are increased, or when alterations are made to heating systems and mechanical equipment, sprinklers must be inspected to determine whether they are appropriate and in good working order. The inspector should make sure that all sprinklers are clean and not painted over, undamaged, and free of corrosion **(Figures 6.53 a and b, p. 196)**. It should also be noted if guards are needed to protect sprinklers against mechanical damage. Carefully examine sprinklers in buildings subject to high temperatures.

Any sprinkler showing evidence of weakness should be replaced with a sprinkler of the proper temperature rating. Weak sprinklers are indicated by a noticeable change in the position of the fusible link (cold flow) or by leakage around the sprinkler orifice. Cold flow is caused by the repeated heating of a sprinkler to near its operating temperature. Cold flow problems can be eliminated either by using a sprinkler with a higher temperature rating or by using frangible bulb sprinklers. Sprinklers exposed to a corrosive atmosphere should have a special protective coating. Replace sprinklers that are corroded, painted, or loaded with foreign material.

Partitions or stock should not obstruct the distribution of water discharge from sprinklers, and the discharge area should be free of hanging displays. A clearance of at least 18 inches (450 mm), which is

Figure 6.51 Inspectors should be appropriately dressed for locations that are likely to be dirty or hazardous.

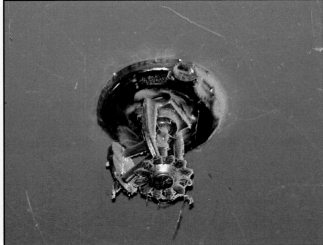

Figure 6.53 a and b Sprinklers need to be inspected to ensure that they are in good condition and free from corrosion.

measured from the deflector to the top of the storage, should be maintained under sprinklers (**Figure 6.54**). However, some sprinklers, such as Large Drop and ESFR sprinklers, require a minimum clearance of 36 inches (900 mm). For this reason it is important to review codes for an individual occupancy *before* the inspection has begun.

A key aspect of the protection provided by an automatic sprinkler system is the continuity of that protection. To enhance system readiness, fire protection standards require that a supply of extra sprinklers be maintained at the protected premises. If sprinklers have been used to control a fire or are damaged, they can quickly be replaced and the system restored to service.

NFPA 13 requires quantities of extra sprinklers as shown in **Table 6.3**. The supply of spares must

Table 6.3
Spare Sprinkler Requirements

Number of System Sprinklers	Spare Sprinklers (Minimum)
1-300	6
301-1,000	12
1,001 or more	24

Reprinted with permission from NFPA 13, *Standard for the Installation of Sprinkler Systems*. Copyright© 2002 National Fire Protection Association, Quincy, MA 02269. This reprinted material is not the complete and official position of the National Fire Protection Association on the referenced subject, which is represented only by the standard in its entirety.

include all types and ratings of sprinklers used in the building. A sprinkler wrench must also be provided.

Sprinkler Piping, Hangers, and Seismic Braces

Inspect all sprinkler piping, hangers and seismic braces to determine that they are in good condition (**Figures 6.55 a and b**). A check should be made for corrosion and physical damage, ensuring that there are no leaks.

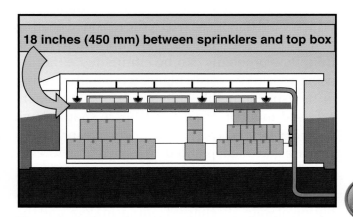

18 inches (450 mm) between sprinklers and top box

Figure 6.54 There should be at least 18 inches (450 mm) of clearance between sprinklers and stored materials. Some sprinklers require more clearance.

Sprinkler piping is *not* to be used as a support for ladders, stock, or other material.

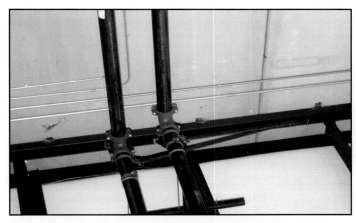

Figure 6.55 a and b Sprinkler hardware needs to be inspected to see that it is in good condition and that there are no leaks.

Water flowing through sprinkler pipes can produce very significant forces, especially in dry-pipe and deluge systems. It is imperative, therefore, that sprinkler piping be properly supported. Improperly supported sprinkler piping and seismic braces will be subjected to stress that can result in breaks and improper drainage. Loose sprinkler hangers and seismic braces should be reported.

Changes in Building Occupancy

A frequent cause of a sprinkler system's inability to control a fire is a change in building occupancy or contents. Many modern sprinkler systems are hydraulically designed; that is, they are designed to supply a given amount of water based on the hazard to be protected. A change in occupancy, however, may result in a hazard greater than the sprinkler system was designed to protect. For example, if a machine shop is converted into a tire warehouse, the change in occupancy will drastically reduce the ability of the sprinkler system to control a fire.

A change in occupancy can be the result of general remodeling as well as a change in contents. If walls are moved, partitions installed, or lighting fixtures relocated, the sprinkler discharge pattern can be obstructed or the number of sprinklers required in a given area may change. During inspections, the inspector should seek answers to the following questions:

- What are the design criteria for the sprinkler system? (This information can be found on the hydraulic nameplate attached to the riser.)
- Has there been a change in occupancy?
- Have combustibles been added that will contribute to a greater fire load or to a more rapid fire spread? **(Figure 6.56)**
- Have alterations caused a need for the reconfiguration of sprinklers?

If the answer to any of these questions is "yes," the inspector will need to note this information on the report and initiate corrective measures immediately. These measures might include immediate remodeling of the system or curtailment of activities in the occupancy until corrections can be made.

Figure 6.56 Fire inspections need to ensure that changes in occupancy also include protection against different fire hazards. A newly opened restaurant will have different fire protection needs than a clothing store.

Inspecting and Testing Wet-Pipe Sprinkler Systems

As previously mentioned, a wet-pipe sprinkler system contains water within the piping at all times. Wet-pipe sprinkler systems are the most reliable of all sprinkler systems. Water under pressure is maintained throughout the system with the exception of the piping to the water motor gong, the main drain piping, and piping from the fire department connection. The system is connected to a water source such as the municipal water system. When a sprinkler fuses, it immediately discharges a continuous flow of water.

When a fire inspector is responsible for performing inspections on wet-pipe sprinkler systems, the primary concerns are four major areas: valves, sprinklers, piping (including hangers and seismic braces), and water supply.

The inspector should ensure that all valves controlling the water supplies to the sprinkler system and valves within the system (sectional valves) are open at all times **(Figure 6.57)**. Any time a valve is found closed, the inspector should report the condition to the responsible agency and the fire department. Examine each control valve for the following:

- Ensure that the valve is opened fully and secured or otherwise supervised in an approved manner; i.e., tamper switches that are chained and padlocked in the open positions, etc. **(Figure 6.58)**.

Figure 6.58 Ensure that valves are secured in the correct positions.

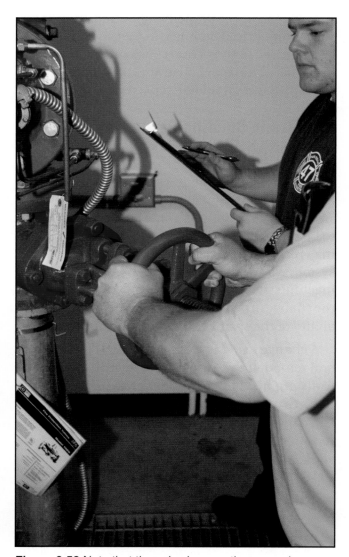

Figure 6.59 Note that the valve is operating properly.

Figure 6.57 Control valves should be locked in the OPEN position.

NOTE: If a valve is closed, do NOT open it until the reason for its closure has been determined. There may be an open pipe in the system that is damaged or under repair. Opening the valve will only cause more damage.

- Check the valve operating wheel or crank to determine its condition **(Figure 6.59)**.

- Make sure that the valve is accessible at all times.

NOTE: If a permanent ladder is provided to elevated valves, check to see that it is in good condition.

- Check the valve operating stem to determine that it is not subjected to mechanical damage. Provide guards if necessary.

- Inspect post indicator valves (PIVs) to ensure that the operating wrench is in place **(Figure 6.60)**. Try the wrench to feel the spring of the rod when the valve is fully opened. The stem should be backed off about one-quarter turn from the fully open position to facilitate ease of operation and to prevent leaks caused by damage to the valve packing.

- Ensure that the target (OPEN/SHUT sign) is readable and that the cover glass is clean and in place.

- Ensure that the PIV bolts are tight and that the barrel casing is intact.

During normal conditions, the inspector should ensure the following:

- The alarm line shutoff cock is completely open.

- Valves to the pressure gauges are open.

- Static pressure above the clapper is greater than or equal to the static pressure below the clapper.

NOTE: Systems without alarm check valves will have only one pressure gauge on the riser.

- Main drain valve, auxiliary drains, and inspector's test valves are closed.

- The retard chamber automatic drip valve should move freely and allow water to drain out of the retard chamber **(Figure 6.61)**.

- The automatic ball drip valve in the fire department connection moves freely and allows trapped water to drain out.

Figure 6.60 Make sure that the operating wrench is in place on post indicator valves (PIVs).

Figure 6.61 Make sure that the retard chamber automatic drip valve moves freely.

- Fire department connection threads are unobstructed, in good condition, and caps are in place.

Piping in wet-pipe systems must be protected against freezing. Windows left open during freezing weather are a frequent cause of frozen piping. Branch lines near these windows can then freeze, leaving the sprinkler system useless. Piping in or near loading docks can also freeze if exposed to freezing conditions, such as occurs when loading dock doors are left open for extended periods. Piping over ceilings in the top floor of a building or in an attic may not receive enough heat during prolonged spells of cold weather.

In addition to stopping the flow of water, frozen piping can also cause failure of control valves and alarm devices. The greatest danger from frozen piping is the rupture of a pipe or its fittings, resulting in severe water damage that requires expensive repairs and interruption of protection. The inspector should be aware of the potential for this problem and be prepared to advise the occupant of preventive measures to eliminate the risk of freezing.

Acceptance Tests

In many jurisdictions, fire department personnel witness sprinkler system acceptance tests. A representative of the installation contractor should conduct these tests. This releases the fire department from liability resulting from damaged equipment because of improper operation or installation. The tests performed and the procedures are as follows:

- *Flushing of underground connections.* Underground mains and lead-in connections should be flushed *before* connection is made to the sprinkler piping. Flushing is continuous until the water is clear and free of debris.

- *Hydrostatic tests.* All piping, including underground piping, is required to be hydrostatically tested at not less than 200 psi (1 400 kPa) for two hours. If the normal static pressure exceeds 150 psi (1 050 kPa), the system should be tested at 50 psi (350 kPa) above the normal static pressure. There should be no visible leakage while the system is pressurized.

NOTE: The IFSTA **Fire Inspection and Code Enforcement** manual contains a variety of forms that may be used for sprinkler system inspections and testing.

Wet-Pipe System Testing

Although in-plant or contract personnel perform and record functional sprinkler system tests at the required intervals, fire department personnel should be familiar with each test. This allows fire personnel to better verify records indicating that the test was performed and the system was found to be operational. Fire department personnel should actually witness these tests.

Testing frequencies for various system components are specified in NFPA 25 and local fire codes and ordinances. In some cases, the occupant's insurance carrier may have more stringent testing requirements than are specified in codes. If the system is equipped with a fire pump, it should be flow tested annually. Diesel engines and electric motors that power fire pumps should be started weekly. (**NOTE:** The procedures for testing fire pumps are covered in Chapter 4.) The water level in gravity tanks should be checked monthly if it is not electrically supervised and quarterly if it is electrically supervised. During cold weather, the heating system for the gravity tank and piping should be inspected daily if it is not electrically supervised and weekly if it is electrically supervised. Following each test, the occupant should file a written record for each system.

When performing functional tests, make sure to notify the central station alarm company that the activation will be a test. The alarm company must be notified again after the system has been restored to a point of readiness. The wet-pipe sprinkler system should be subjected to two main types of tests: the waterflow alarm test and the main drain test.

Waterflow alarm test. Alarms should be tested quarterly. The inspector's test valve should only be used during nonfreezing weather to avoid ice formation on sidewalks and roadways. In addition, the alarm piping can be damaged or obstructed by ice formation. During tests, all components should be visually checked for obvious problems.

On a wet-pipe system with an alarm check valve, the inspector should witness the opening of the alarm bypass valve to test the alarm without unseating the valve **(Figure 6.62)**. The pressure gauge readings should not change significantly, but water should flow to the retard chamber (if so equipped) and then to the alarm line. The water motor gong or electric alarm should sound. The retard chamber drain should empty the chamber after the alarm bypass valve is closed. If there is no retard chamber, the alarm line should be drained at the conclusion of the test.

On all types of wet systems, a waterflow alarm test should be conducted by using the inspector's test valve. In accordance with NFPA 25, vane-type water flow switches are required to be tested semiannually. Pressure switches and water motor gongs are required to be tested quarterly. The inspector's test valve simulates the operation of a single sprinkler and ensures that the alarm will operate even if only one sprinkler is fused in a fire **(Figure 6.63)**.

In older systems, the inspector's test valve was usually located at the end of the most remote branch line. In newer systems, the inspector's test valve may be located near the sprinkler riser so that one person is capable of performing the test. The test valve consists of a 1-inch (25 mm) pipe equipped with a shutoff valve and a discharge orifice equal in size to the smallest sprinkler in the system.

With an observer at the riser, another individual opens the inspector's test valve. The alarm should sound and only a slight variation in pressure should be observed at the riser. After the alarm operates, the inspector's test valve is then manually closed.

Main drain test. According to NFPA 25, the main drain test should be conducted annually. The fire department inspector should visually inspect the system and witness the alarm test in conjunction with the main drain test.

Each sprinkler system riser has a main drain. Risers that are 4 inches (100 mm) or larger in diameter will be equipped with a 2-inch (50 mm) main drain **(Figure 6.64, p. 202)**. The primary purpose of the main drain is to simply drain water from the system for maintenance purposes. Because a

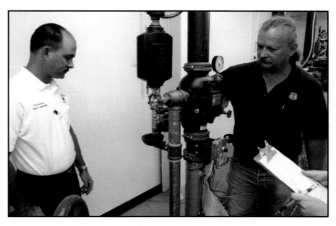

Figure 6.62 The alarm bypass valve should be opened to test the alarm.

Figure 6.63 The waterflow alarm test should be conducted by using the inspector's test valve.

large volume of water flows when the main drain is opened, however, it can also be used to check the system water supply.

The main drain test is useful for detecting such impairments as closed valves, obstructions, or gradual deterioration in the water supply. To a limited degree, it can also be used as an indicator of the overall water supply in the area. To perform a main drain test, the inspector should witness the building representative performing the following steps:

Step 1: Observe and record the pressure on the gauge at the system riser (**Figure 6.65**).

Step 2: Watch as a building representative fully opens the main drain.

NOTE: The main drain usually discharges outside the building, and the area should be checked to make sure it is clear (**Figure 6.66**).

Step 3: Observe and record the pressure drop.

Step 4: Observe the building representative closing the 2-inch (50 mm) main drain slowly (**Figure 6.67**).

NOTE: It is extremely important to close the valve very slowly in this case to avoid a water hammer resulting in an erroneously high static pressure.

Step 5: Observe and record the final static pressure. If it is not the same as the initial static pressure, it is likely there was trapped pressure in the system.

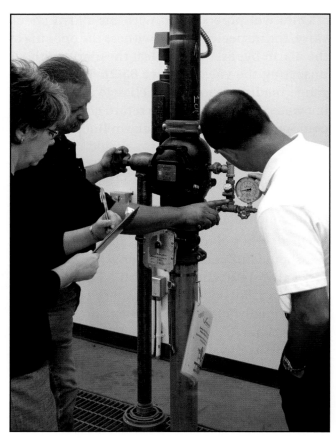

Figure 6.65 Record the pressure on the gauge at the system riser.

Figure 6.64 The main drain on a riser.

Figure 6.66 Check the area around the main drain to make sure it is clear.

Figure 6.67 The main drain must be closed slowly.

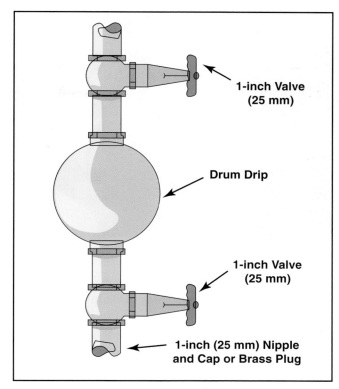

Figure 6.68 A drum drip valve.

These readings should be compared to previously recorded readings. If significant differences are noted, a supply valve may be partially closed or there may be an obstruction in the supply line. On a system using an alarm check valve, the pressure readings should be taken from the lower gauge because erroneously high static pressures can exist above the valve.

Inspecting and Testing Dry-Pipe Sprinkler Systems

Dry-pipe systems have many elements in common with wet-pipe systems and they are inspected in the same manner as wet systems. However, they also have unique features that require special attention. During an inspection of a dry-pipe sprinkler system, inspectors should ensure the following:

- All indicating control valves are open and properly supervised in the open position.
- Air pressure readings correspond to previously recorded readings.
- The ball-drip valve moves freely and allows trapped water to seep out of the fire department connection.
- The velocity drip valve located beneath the intermediate chamber is free to move and allow trapped water to seep out. Inspectors can check this valve by instructing someone to lift a push rod that extends through the drip valve opening.

Where an automatic drip valve is installed, the velocity drip valve can be checked by having someone move the push rod that is located in the valve opening.

- The fire department connection threads are unobstructed, in good condition, and caps are in place with gaskets intact. Identification signs must be in place and readable.
- Any drum drips are drained to eliminate the moisture trapped in the low areas of the system (**Figure 6.68**). This should be done frequently enough to prevent clogging. The schedule will vary depending on the environmental conditions in the area.
- During freezing weather, the dry-pipe valve enclosure heating device keeps the temperature of the dry-pipe valve at or above 40° F (4° C).
- The priming water is at the correct level (**Figure 6.69, p. 204**). If necessary, personnel can drain water by opening the priming water test level valve until air begins to escape.

NOTE: If the system is equipped with a quick-opening device, opening the priming water test line could trip the system.

Figure 6.69 The priming water must be at the correct level.

- The system's air pressure is maintained at 15 to 20 psi (105 kPa to 140 kPa) above the trip point and no air leaks are indicated by a rapid or steady air loss. If inspectors note excessive air pressure, they should have the system drained down.

- The system air compressor is approved for sprinkler system use, well maintained, operable, and of sufficient size.

Acceptance Tests

As with wet-pipe sprinkler systems, acceptance tests for dry-pipe systems are conducted by a representative of the installation firm, but they should also be witnessed by fire department personnel. The tests performed and the procedures for each are as follows:

- *Flushing underground connections.* Underground connections should be flushed following the same procedures described for wet-pipe sprinkler systems.

- *Hydrostatic testing.* Hydrostatic testing of dry-pipe systems is performed in the same manner as described for wet-pipe sprinkler systems. In freezing weather, however, dry-pipe systems are tested for 24 hours with not less than 40 psi (280 kPa) air pressure. If there is a loss of more than 1½ psi (10 kPa), the leaks should be located and corrected. When the weather warms, the system should be subjected to a hydrostatic test using water.

Dry-Pipe System Testing

Fire personnel should encourage facility personnel to visually inspect dry-pipe sprinkler systems weekly. A weekly inspection is recommended because regular inspections enable building personnel to become familiar with the system and help them to discover deficiencies early. In the case of dry-pipe systems, it is particularly important to check the system air pressure on a weekly basis. According to NFPA 25, heat for the dry-valve enclosure should be checked daily during cold weather. The main drain and alarm tests should be performed annually, but not during freezing weather to avoid piping damage and ice accumulations.

Low-point drains should be tested each fall. A trip test of the dry-pipe valve should be made annually with the main control valve partially open. A full flow trip test is recommended every three years. According to NFPA 25, quick-opening devices should be tested quarterly.

Two types of tests, the main drain test and the trip test, are recommended for dry-pipe sprinkler systems. The following sections detail the procedures for each of these tests.

Main drain test. The main drain test is conducted on dry-pipe systems for the same reasons it is conducted for wet-pipe systems. The same four-step procedure is used for both systems and is described earlier under the wet-pipe system test section.

Trip test. The operational test of a dry-pipe valve is known as a "trip test," or sometimes as a "full flow trip test." A dry-pipe valve should be full flow trip tested every three years to ensure that the air seat has not become stuck and that the clapper valves will move freely. During alternate years the trip test should be conducted with the valve partially (slightly) opened to prevent water from flowing into the sprinkler piping. The trip test of a dry-pipe valve permits water to flow into the system.

When planning a trip test, the inspector should be aware that this procedure can take from two to four hours to complete. The length of time needed depends on the amount and size of the piping in the system and the capacity of the air compressor to pressurize the system. Older valves may prove more difficult to reset because of leaking gasket seats, and more than one attempt may be required. The inspector should use the following steps when observing the performance of a trip test:

NOTE: The building representative should perform the actual steps and the inspector should verify that the system and component parts are working properly.

Step 1: Have a building representative open the main drain fully. This action flushes any sediment or scale that may be in the water supply and prevents it from entering the sprinkler piping. Slowly close the drain valve.

Step 2: Check the system water control valve for freedom of movement by slightly opening the main drain and turn the water control handwheel or crank (**Figure 6.70**). The reason for opening the main drain at this point is to prevent accidental tripping of the valve. Closing the water control valve without opening the drain may "squeeze"

the water trapped between the control valve and dry-pipe clapper. Although water is very slightly compressible, it is considered incompressible for fire department purposes; therefore, something has to give. In some cases, the clapper will be pushed off its seat.

Step 3: Leave the water control valve partially open and have the building representative close the main drain. Having the valve only partially open permits more rapid closing when the dry-pipe valve trips, thus minimizing the total amount of water flow into the system.

Step 4: Bleed air from the system by having the building representative open the primary water valve or the valve body drain if the valve is so equipped (**Figure 6.71**). This should be performed while observing the reading on the air-pressure gauge.

Step 5: Record the tripping point of the dry-pipe valve as indicated by the air-pressure gauge, and close the control valve.

Step 6: Check water and air-pressure gauges to ensure that pressure equalization has occurred after tripping (**Figure 6.72, p. 206**).

Step 7: Verify that the local alarm and control panel, central station alarm, or fire department alarm has operated.

Figure 6.70 The main drain must be opened to prevent accidental tripping of the valve.

Figure 6.71 Observe the reading on the air pressure gauge as air is bled from the system.

Step 8: Have the building representative open the main drain valve (**Figure 6.73**). Be sure that the system is completely drained before proceeding. A considerable amount of water will accumulate in the system riser, and it will rush out when the valve cover is removed if not properly drained.

Step 9: Remove the dry-pipe valve cover. The inspector should check to ensure that the clapper is latched in the open position (**Figure 6.74**).

Step 10: Clean the air and water clapper seats and remove any debris from the valve housing (**Figure 6.75**). Check the condition of the air seat (this part is usually made of rubber). Check for any signs of impurity such as oil in the water.

Step 11: The building representative releases the clapper latch and reseats the valve.

Step 12: The building representative replaces the valve cover (**Figure 6.76**).

Step 13: The building representative adds priming water through the priming water fill pipe.

Figure 6.73 After the main drain valve is opened, make sure that the system is completely drained before proceeding.

Figure 6.74 Make sure that the clapper is latched in the open position.

Figure 6.72 The water and air-pressure gauges should be equalized.

Step 14: The building representative closes all drain valves.

Step 15: Once all drain valves are closed, the building representative pressurizes the system with air or nitrogen to the proper pressure (**Figure 6.77**).

Step 16: Check for a flow of water at the intermediate chamber drain. It is not unusual for water to drip out initially, but a steady stream of water usually indicates that the air seat is not properly seated.

Step 17: The building representative opens the main drain partially and then slowly opens the water control valve fully.

Step 18: The building representative slowly closes the main drain.

Step 19: Check the air- and water-pressure gauges. The air pressure should be lower than the water pressure. If the two gauges read the same, it is an indication that the valve has tripped, and steps 9 through 18 must be repeated.

Step 20: The building representative notifies the alarm service that work has been completed.

Step 21: Attach to the valve a tag that shows the date the valve was tested, the air pressure at which the valve tripped, and the name of the person who performed the test (**Figure 6.78**).

Figure 6.75 Remove any debris that may have accumulated in the valve housing.

Figure 6.77 The building representative pressurizes the system with air or nitrogen.

Figure 6.76 Watch as the valve cover is replaced.

Figure 6.78 The valve should be tagged with information about the test.

Inspecting and Testing Deluge and Preaction Sprinkler Systems

Deluge and preaction system inspections follow the same guidelines as those performed on wet and dry systems with regard to piping and valves. These systems should be visually inspected monthly. Weekly or daily inspections are also recommended depending on certain conditions such as cold weather. During cold weather, the sprinkler system enclosure should be checked daily. The supervisory air pressure in a preaction system should be checked weekly. The detection system should be tested semiannually. The system itself should be trip tested annually. A written record of all tests and inspections should be made, compared to earlier tests, and filed.

Acceptance testing for deluge and preaction sprinkler systems should follow the same procedures listed for dry-pipe systems. Deluge and preaction systems should have main drain, alarm, and trip tests performed. The following sections highlight the procedures for each of these tests.

Main Drain and Alarm Tests

The main drain test for deluge and preaction sprinkler systems serves the same purpose as it does on wet- and dry-pipe systems. It is conducted as previously described. Alarms on deluge and preaction systems may be somewhat more complicated than on wet or dry systems because they are usually designed to protect special hazards. However, some deluge systems make use of ordinary alarm lines that can be tested by opening an alarm bypass valve (**Figure 6.79**).

Trip test. The following procedure details the process for trip-testing deluge and preaction sprinkler systems. As with dry-pipe systems, the inspector should be aware that this test is time-consuming.

Step 1: The inspector should notify the supervising agency of the sprinkler system, such as a central alarm service or fire department, that a test is to be performed.

Step 2: The building representative replaces the open sprinklers on a small deluge system with standard sprinklers or plugs (**Figure 6.80**). On large systems where the sprinklers cannot be replaced or plugged, the main water control valve may be closed to within two turns from closed. This procedure permits rapid shutoff of the water when the valve trips.

Figure 6.80 Replace the open deluge sprinklers with regular sprinklers or plugs.

Figure 6.79 Some alarm lines can be tested by opening the alarm bypass valve.

NOTE: Remember, in a deluge system all sprinklers are open.

Step 3: The building representative activates the system by activating a heat or smoke detector, as the case may be, or by using the manual trip (**Figure 6.81**).

Step 4: After the valve trips, the building representative immediately closes the main water control valve and drains the system using the main drain. The nozzles of water spray fixed systems and foam deluge systems must be checked to ensure that nozzles are not plugged, the distribution is not obstructed, and that the surface being protected is properly wetted. This must be determined before the control valve is closed.

Step 5: The building representative opens the valve cover, cleans the valve seat, and removes any debris from the valve housing.

Step 6: The building representative unlatches the clapper and resets the valve (**Figure 6.82**).

NOTE: Some deluge and preaction valves do not require resetting by removal of the valve cover. Refer to manufacturer's instructions for proper resetting procedures.

Step 7: The building representative reinstalls the cover (**Figure 6.83**).

Step 8: If the pipe integrity is supervised, the building representative pressurizes the system to the proper air pressure.

Step 9: The building representative restores water pressure by opening the supply control valve with the main drain open. When the supply valve is completely open, the building representative slowly closes the main drain and checks the gauges.

Step 10: The building representative removes plugs or caps from sprinklers on deluge systems or reinstalls open sprinklers.

Figure 6.81 The building representative activates the system.

Figure 6.82 The building representative unlatches the cover and reseats the valve.

Figure 6.83 The cover is reinstalled.

Restoring Sprinkler Systems

It is most desirable for the sprinkler system to be fully restored to service before the fire department's departure from the premises. Unfortunately, the appropriate maintenance and sprinkler service personnel are not always immediately available. Even when personnel are available, the process for restoring dry-pipe, deluge, and preaction systems can take several hours. Each incident will be different. Any one of the following actions may be dictated in fire department standard operating procedures for system restoration:

- For safety reasons when practical, one fire company may be left on the scene to verify restoration of the system.

- A chief officer may continue to recheck the occupancy until the system is restored.

- Responsibility for the building's welfare may be turned over to the occupant or a responsible agent of the occupant. (It is good business practice for fire department personnel to make a follow-up call to verify system restoration).

Sprinklers that have activated must be replaced with identical sprinklers if at all possible. It is very easy to replace a sprinkler with one that "looks" the same. The difference between sprinklers having different RTIs, orifice sizes, K-factors, and temperature ratings is very subtle. The best way to ensure proper placement is to look for the Sprinkler Identification Number (SIN) for newer sprinklers. Older sprinkler sprinklers do not use the SIN designation. The SIN is specific to the make and model of the sprinkler and is stamped on the deflector. Where there is no SIN, care must be taken to ensure that the same RTI, orifice size, K-factor, and temperature rating are used. A sprinkler contractor or a fire protection authority should check for compliance and fire department staff should conduct a follow-up inspection.

The fundamental responsibility for restoration of a system belongs with the building owner. For the same reasons of liability explained earlier, fire department personnel should not perform restoration or any other hands-on procedures on sprinkler systems. NFPA 13 requires that the sprinkler contractor provide the owner with instructions in the proper operation and maintenance of the system and with a copy of NFPA 25. Although fire department personnel will not be required to perform the restoration, it is good for them to understand what is required to return a system to use. The following sections highlight these procedures.

Restoring Wet-Pipe Sprinkler Systems

Restoring a wet-pipe system that has operated during a fire is usually a simple matter. Alarm valves are self-restoring and require no special procedure other than shutting off the water to replace fused sprinklers. It may be necessary to silence and reset the alarms, depending on the type of alarm used. Water motor gongs do not require special attention; however, the retard chamber should be checked to see that it has drained. The alarm will stop ringing when the alarm valve resets.

Restoring Dry-Pipe Sprinkler Systems

Unlike wet-pipe systems, dry-pipe systems do not reset themselves. The dry-pipe valve must be opened and reset manually. The procedure for resetting a dry-pipe valve is described in detail in the section on dry-pipe system trip testing. It is a fairly complex procedure that may prove difficult on older valves and can take considerable time.

Restoring Deluge and Preaction Sprinkler Systems

Deluge and preaction systems make use of a variety of valves and detection systems. The exact procedure for restoring a system depends on the specific equipment used. Some detection devices reset automatically; others need to be replaced. Some deluge valve fixtures must be opened up to reset the valve.

Sprinkler System Impairment Control

To provide protection, sprinkler systems must be in service constantly. Interruptions in service can occur, however. These interruptions can be either planned or unplanned. Planned interruptions include shutting down the system for maintenance, testing, or building renovation. Unplanned interruptions occur from such causes as frozen pipes, broken pipes, and equipment failures.

It is critical that additional fire safety measures be applied when sprinkler systems are inoperative. These measures may include posting fire guards, suspending hazardous operations, and adding manual equipment such as hoselines. There are times when the fire department may be requested to assist in providing support during a planned or unplanned impairment to a fire protection system. Whenever a sprinkler system is taken out of service, the responsible party should notify the fire department even if the period of interruption is brief. Costly fires have occurred during the short period that a system was out of service.

NOTE: NFPA 25 states that whenever a sprinkler system or underground system has been taken out of service, a tag must be placed on the FDC. In addition, the supervising station and the fire department must be notified or if local codes require it, the fire marshal.

Residential Sprinkler Systems

According to records kept by the NFPA, approximately 80 percent of fire fatalities occur in residential occupancies **(Figure 6.84)**. It is logical, therefore, to extend the protection provided by automatic sprinkler systems into the residential environment. Within the last 20 years, a very deliberate effort has been made to provide automatic sprinkler protection in residential buildings. The United States Fire Administration (USFA) funded research on the subject, resulting in two residential sprinkler standards written by NFPA: NFPA 13D, *Standard for the Installation of Sprinkler Systems in One- and Two-Family Dwellings and Manufactured Homes,* and NFPA 13R, *Standard for the Installation of Sprinkler Systems in Residential Occupancies Up To and Including Four Stories in Height.* These standards were developed so that systems could be installed economically in residential buildings with the goal of maintaining survivable conditions in the structure during an outbreak of fire. The survivable conditions were considered to be the following:

- Maximum gas temperature at eye level 200°F (93°C)
- Maximum ceiling surface temperature 500°F (260°C)
- Maximum carbon monoxide concentration 1,500 parts per million (ppm)

Figure 6.84 Fire fatalities are most likely to occur in residences. *Courtesy of District Chief Chris E. Mickal.*

There are several barriers to the application of conventional sprinkler technology to residential buildings, especially single-family dwellings:

- The hardware of standard sprinkler systems, such as fire department connections and alarm valves, is large and obtrusive if applied to residential applications under the same rules as used in commercial and industrial applications.

- Conventional industrial systems, if applied to ordinary dwellings, would be beyond the economic means of a large segment of the population.

- Water supply requirements are substantially reduced for residential systems. A minimum 10-minute supply is required for systems designed in accordance with NFPA 13D and a minimum 30-minute supply is required for systems designed in accordance with NFPA 13R.

To make sprinklers useful in residential applications, changes in design were made. These changes were needed not only to decrease the cost of the systems but also to enhance their effectiveness in protection of life. It must be remembered that the original automatic sprinkler technology was developed primarily to protect industrial property, not human life. To reduce the loss of life in residential fires, somewhat different design criteria were necessary and appropriate. Modifications included changes in the following areas:

- Residential sprinklers were designed to be fast response.

- Water supply requirements were reduced to levels appropriate for life safety.

- Areas of coverage for sprinklers were adjusted based on sprinkler design and typical residential fire loads.

- Alarms were made simpler and more realistic for residential applications and smoke detectors were approved for notification.

- Valve arrangements were made so that they would be unobtrusive in a common residence (**Figure 6.85**).

- Sprinklers were designed to discharge water high on the walls.

Figure 6.85 Residential sprinkler valves can be placed in unobtrusive locations.

Residential Sprinklers Versus Conventional Sprinklers

Residential sprinklers differ from those used in commercial systems and are tested under a different standard by Underwriters Laboratories. A major difference between residential sprinklers and commercial sprinklers is their sensitivity or speed of operation. To satisfy the previously stated criteria for maintaining a survivable atmosphere, residential sprinklers must operate more quickly than commercial sprinklers. Although a commercial sprinkler may have an operating temperature of 165°F (74°C), the thermal lag of the fusible link may delay the activation of the sprinkler until surrounding air temperature is considerably higher. By redesigning the fusible link, the sprinkler can be made to activate before conditions in the room become lethal for the occupants.

Figure 6.86 Residential sprinklers are installed high on the walls to prevent fire from traveling above the sprinkler spray.

Residential sprinklers also have distribution patterns that are different from commercial sprinklers. Residential sprinklers are designed to discharge water higher on the enclosing walls of a room. This is to prevent a fire from traveling above the spray as might occur with burning drapes or in pre-flashover conditions **(Figure 6.86)**.

Residential Sprinkler Piping

Any type of piping, such as plastic that is listed by an approved testing agency, may be used for residential sprinkler systems. Steel and copper may be used but are not required to be listed. The minimum pipe size that may be used in a residential system is ¾-inch (19 mm) for plastic and copper piping and 1-inch (25 mm) for steel piping.

Water Supply and Flow Rate Requirements

The water supply requirements for residential sprinklers are less than those for commercial systems. NFPA 13D requires only 18 gpm (72 L/min) for any single sprinkler. If there are two or more sprinklers, each requires 13 gpm (52 L/min) as a maximum required water supply. In addition, the water supply needs only to supply this flow rate for 10 minutes.

For larger multiple dwellings, the designed flow rates are greater. NFPA 13R requires 18 gpm (72 L/min) for a single sprinkler and not less than 13 gpm (52 L/min) to a maximum of *four* sprinklers. The water supply for these larger buildings is required to supply the sprinklers for 30 minutes.

In either case, the listed flow rates for sprinklers may be used. These flow rates are typically less than those specified in NFPA 13D and NFPA 13R. The flow rates are based on a minimum density of 0.05 gpm/ft² (20.3 L/min/m²).

The reduction in water supply is prompted not only by economics but also by the fact that many residential buildings are served only by small domestic supplies. Many single-family dwellings, for example, are supplied by wells. Although a 10-minute supply may not completely control all fires, it will prevent room flashover and give occupants an opportunity to escape.

The water supply for residential sprinklers may be taken from several sources. These sources may include a connection to the public water system, an on-site pressure tank, or a tank with an automatic pump. A connection to a public water system is reliable and usually provides adequate volume, pressure, and duration; however, rural homes may not be serviced by public water systems. **Figure 6.87**

Figure 6.87 Some residences may need a tank to supply water for the sprinkler system.

illustrates a tank with an attached automatic pump used to supply a single-family dwelling sprinkler system. Notice that this equipment is much more compact than industrial equipment.

To be of value, a residential sprinkler system, like any other system, must be continually in service. As in a more conventional sprinkler system, inadvertent or deliberate closing of valves renders the system useless. When a residential sprinkler system is supplied from a public water system, the possibility of the supply valve being closed can be virtually eliminated by using one valve arranged to control both the sprinkler system and the domestic water supply. With this arrangement, the sprinklers cannot be shut off without shutting off water supply to the household. Where plumbing or water department requirements will not permit this type of uninterrupted connection, NFPA 13D permits the sprinkler valve to be supervised or simply locked in the open position.

Residential Sprinkler Spacing

Sprinkler coverage in residential systems is not as extensive as in conventional sprinkler systems. In single-family dwellings, NFPA 13R and 13D allow sprinklers to be omitted from many areas, includ-ing bathrooms, small closets, garages, porches, carports, uninhabited attics, and entrance hallways. Many of these areas may be omitted only if certain size and construction requirements are met. A layout of a residential sprinkler system is shown in **Figure 6.88**.

The basic spacing for sprinklers in residential systems is a maximum of 144 square feet (13.4 m²) per sprinkler. The maximum spacing between sprinklers is 12 feet (4 m) with the maximum allowable distance of a sprinkler from a wall being 6 feet (2 m). However, sprinkler manufacturers have produced a variety of sprinkler designs, and the spacing of sprinklers may be based on the spacing for the particular sprinklers that have been tested and listed. Some residential sprinklers, therefore, can be spaced to protect an area as large as 20 feet × 20 feet (6.1 m × 6.1 m). However, with this sprinkler spacing, the minimum discharge per sprinkler is increased and will vary according to the make and model of the sprinkler. For example, one manufacturer requires 28 gpm (112 L/min) for a single sprinkler and 24 gpm (96 L/min) for multiple sprinklers.

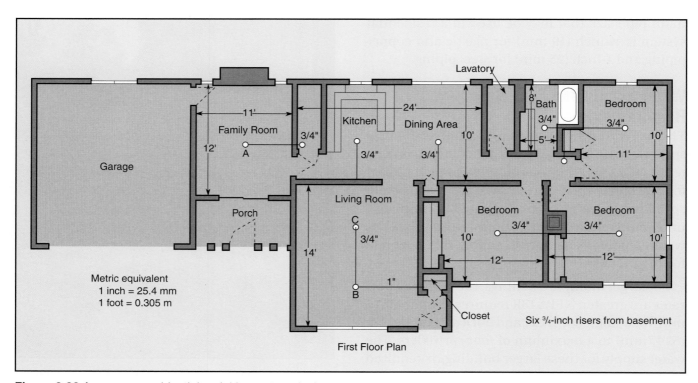

Figure 6.88 A common residential sprinkler system design.

NOTE: For more information about fire department operations at sprinklered occupancies, see **Appendix E**.

Summary

Automatic sprinkler systems, along with other water-based fire suppression systems, provide an essential life and property conservation service and are immediately available when a fire occurs in a structure where they are installed. Numerous lives and countless dollars have been saved since sprinkler systems were first introduced over 100 years ago. This chapter has presented an overview of modern automatic fire sprinkler application and technology as well as descriptions of the standards and codes that govern these applications. All basic components of sprinkler systems have been described in general terms so that the inspector/firefighter will possess the knowledge required to properly identify each of these components and their operation.

It is imperative that fire service personnel possess a good working knowledge of sprinkler system piping systems, pipe components, and operating methods of the different types of valves that are utilized throughout an automatic fire sprinkler system.

Knowledge of the basic types of sprinklers: wet-pipe, dry-pipe, deluge, and preaction is also a must, as well as familiarity with the typical structures, circumstances, and commodities for which each of these system types would be employed. The various national codes and standards are an integral part of the knowledge necessary for determining the optimum system for a given application.

To ensure the reliability of the automatic fire sprinkler system and its operational readiness, methods of inspecting and testing the system have been described for use by the inspector. Emphasis has been placed on those critical procedures that must be performed during a periodic inspection. Also included are the processes and procedures that are required when an automatic fire sprinkler system is being restored to service.

The design and operation of residential fire sprinkler system is another area in which the firefighter/inspector must be knowledgeable as these systems become more prevalent. Differences between these systems and those for automatic fire sprinklers in commercial or industrial applications must be known if minimum design requirements needed to provide a level of life safety in one- and two-family residences are met.

Special Extinguishing Systems

1. Describe the major distinctions between an automatic sprinkler system and a specialized extinguishing system.

2. Describe the hazards for which a wet chemical system is best suited.

3. Name the major components of a wet chemical system that that should be inspected for damage and operability

4. Name at least four hazard areas for which dry chemical extinguishing systems are well suited.

5. Name at least four hazard areas for which gaseous extinguishing systems are well suited.

6. Describe the most serious safety problem associated with the use of carbon dioxide systems and discuss how to minimize it.

7. Describe typical areas in which foam extinguishing systems are likely to be installed.

8. Discuss the importance of foam concentrate, foam proportioner, foam solution, and finished foam.

9. Describe the venturi principle as it applies to the formation of foam solution.

10. Describe the basic types of foam proportioners in use.

11. Identify various types of foam nozzles and their applications.

FESHE Objectives

**Fire and Emergency Services Higher Education (FESHE) Objectives:
Fire Science Curriculum: Fire Protection Systems**

• Comprehend types, components, and operation of automatic, special sprinkler systems, and standpipes.

Chapter 7
Special Extinguishing Systems

Special extinguishing systems are used where automatic sprinklers may not be the best solution to fire problems. These locations can contain food preparation equipment, combustible metals such as magnesium or potassium, or highly sensitive computer and electronic equipment **(Figure 7.1)**. A critical feature of most special extinguishing systems is that they have a limited amount of extinguishing agent, unlike automatic sprinkler systems, which are supplied by a nearly endless amount of water. Another important distinction between the two systems is that an automatic sprinkler system is considered successful when it *controls* a fire, whereas a specialized extinguishing system must *extinguish* the fire to be successful. This chapter addresses the following specialized extinguishing systems:

- Wet chemical
- Dry chemical
- Gaseous agents
- Foam

 It is important to note that no single extinguishing agent is suitable for all systems.

Wet Chemical Extinguishing Systems

A wet chemical system is best suited for applications in commercial cooking hoods, plenums, ducts, and associated cooking appliances. This type of chemical extinguishing system is used in situations where rapid fire knockdown is required and where reignition is unlikely.

Figure 7.1 Many areas require an extinguishing system that does not discharge water onto sensitive equipment. *Courtesy of Tom Jenkins.*

The wet chemical agent is typically a solution of water and either potassium carbonate or potassium acetate that is delivered to the hazard area in the form of a fine spray. It is an effective extinguishing agent for fires involving such flammable liquids as grease or oil or ordinary combustibles such as paper and wood **(Figure 7.2, p. 220)**. The nature of the chemical is such that it reacts with animal or vegetable oils and forms a soapy, foam-like film. This reaction is called *saponification*. The agents used in this system are the same as those described for portable fire extinguishers in Chapter 1.

NOTE: Refer to NFPA 17A, *Standard for Wet Chemical Extinguishing Systems,* 2002 edition, for more information about wet chemical systems.

Wet chemical systems can be messy, particularly when food greases are involved. The spray from a wet chemical system can also migrate to surround-

Figure 7.2 Wet-chemical agents form a soapy film that helps extinguish fires in cooking oils.

ing surfaces, causing corrosion of electrical wires or damage to hot cooking equipment. For this reason, it is very important to ensure a prompt cleanup after activation of a wet chemical system. This type of system is not recommended for electrical fires because the spray may act as a conductor.

System Design

A wet chemical system can be grouped into two broad design categories: engineered and pre-engineered. An engineered system is designed and constructed for a specific hazard; it is also likely to be a large, expensive, custom-designed system. The most common system, the pre-engineered or package system, is calculated to protect areas of a given size. Many pre-engineered systems are installed in commercial cooking areas and in hood/duct systems.

Essentially, all wet chemical systems have the following components:

- Storage tank for expellant gas and agent
- Piping to carry the gas and agent
- Nozzles to disperse the agent
- Actuating mechanism

There are no standard container sizes for agents. The storage container may contain both the agent and the pressurized expellant gas (stored pressure), or the agent and the gas may be stored separately **(Figure 7.3)**. A pressure gauge attached to the container is an indication of a stored-pressure

container. The expellant gas is either nitrogen or carbon dioxide. The containers are basically the same as those described for portable fire extinguishers except that the system containers are much larger. The tanks must be located as close to the discharge point as possible, but they should also be in an area with a temperature range of -40°F to 120°F (-40°C to 49°C).

The wet chemical agent is delivered to the hazard through nozzles attached to a system of fixed piping **(Figure 7.4)**. The piping is specially designed to account for the unique flow characteristics of the

Figure 7.3 This extinguishing agent and pressurized gas are stored separately.

Figure 7.4 These nozzles and piping are used to deliver the wet chemical agent.

agent. The proper size, the number of bends and fittings, and the pressure drop (friction loss) are calculated into piping requirements.

Wet chemical is released into the piping system in response to activation devices. These devices are usually designed to activate when fusible links melt in response to heat (**Figure 7.5**). The fusible links trigger a mechanical or electrical release that in turn starts the flow of expellant gas and agent. According to NFPA 17, fixed systems must be capable of manual activation and must be equipped with automatic fuel or power shutoff (**Figure 7.6**). The shutoff device must be restored manually. Detailed maintenance and restoration of these systems should be left to trained personnel. Trained personnel should also be able to inspect these systems for the following:

- Mechanical damage
- Aim of nozzles
- Change in hazard
- Proper pressures on stored-pressure containers
- Maintenance tag for scheduled service

Inspection and Test Procedures

NFPA 17A recommends that the following procedures be performed monthly to ensure system readiness:

- Check all parts of the system to ensure that they are in their correct location and connected properly.
- Inspect manual actuators for obstructions.
- Ensure that tamper indicators and seals are intact (**Figure 7.7**).
- Determine that the maintenance tag or certificate is placed properly and contains correct information.
- Examine the system for obvious damage.
- Check that pressure gauges read within their operable ranges.
- Check nozzle blowoff caps, if provided, for damage.
- Examine the area to ensure that the protected equipment or hazard has not been modified, replaced, or relocated.

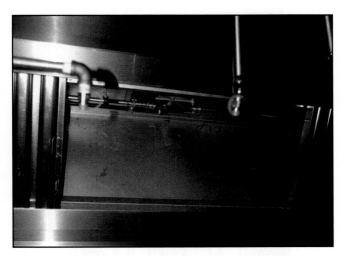

Figure 7.5 An example of a fusible link in a kitchen hood.

Figure 7.6 Manual activation for a fixed system.

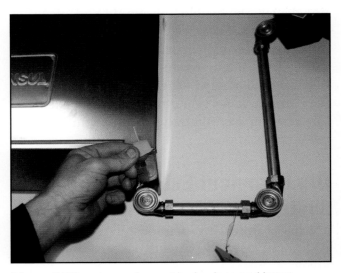

Figure 7.7 Tamper seals must be in place and intact.

A record should be kept of each inspection and any problems noted should be corrected immediately. In most cases, this will require notification of the system service provider. NFPA 17A recommends a more detailed inspection on a semiannual basis. The following procedures, to be performed by a trained individual, are recommended for this examination:

- Ensure that the nature of the hazard being protected or the size of the hazard area has not changed.

- Examine all components thoroughly.

- Check nonpressurized containers to see that liquid levels are correct.

- Make sure that the piping is not obstructed.

- Examine the system to determine that there is no evidence of corrosion, structural damage, or repairs (soldering, welding, etc).

- Test all working components in accordance with the manufacturer's instructions.

  NOTE: Normally, discharging the agent would not be required at this time.

- Correct any problems noted in accordance with the manufacturer's recommendations.

- Ensure that the maintenance report and recommendations are forwarded to a responsible person.

- Note that the maintenance tag is in place with current information clearly indicated.

If the system actuator is controlled by a fusible link device, the fusible link should be replaced at least annually. If it appears to be distorted from frequent exposure to heat, it may need to be replaced more often. If the system actuator is controlled by sensing elements other than fusible metal, it may remain in service provided it is cleaned and inspected in accordance with the manufacturer's recommendations.

Dry Chemical Extinguishing Systems

The dry chemical fire extinguishing system is most commonly used to protect the following hazard areas:

- Flammable and combustible liquids and gases
- Dip tanks
- Paint spray booths
- Exhaust duct systems
- Commercial cooking equipment
- Transformers
- Heavy diesel equipment such as earthmovers, etc.

All dry chemical systems should meet the requirements set forth in NFPA 17, *Standard for Dry Chemical Extinguishing Systems.* There are two main types of dry chemical systems: local application and total flooding **(Figure 7.8)**. The local application system, which is the most com-

Figure 7.8 A total flooding dry chemical system.

mon type of dry chemical system, is designed to discharge agent onto the target hazard. A total flooding system is designed to introduce a thick concentration of agent into a closed area such as a paint spray booth.

All dry chemical agents discharge a cloud of chemical that leaves a residue. This residue creates a cleanup problem after system operation. The powder can usually be vacuumed; however, the very fine particle size makes it somewhat difficult to contain in an ordinary vacuum bag. Because the chemicals are water repellant, normal mopping is also difficult. Due to these cleanup problems, dry chemicals are not considered "clean" agents. A dry chemical system is not recommended for an area that contains sensitive electronic equipment. The chemical residue has insulating characteristics that hinder the operation of the equipment unless extensive restorative cleanup is performed. Dry chemical systems are not suitable for fires involving flammable metals or fires that have deep-seated burning activity. Dry chemical systems are effective on such electrical components as oil-filled transformers or circuit breakers.

The components and actuation of dry chemical systems are virtually the same as those for wet chemical systems. The inspection and testing procedures are also the same with a few exceptions. See NFPA 17 for a full description of requirements.

Gaseous Systems

Gaseous fire extinguishing systems are used primarily where wet or dry extinguishing systems may not be desirable or suitable. The agents and their properties discussed in this section are carbon dioxide (CO_2), halogenated agents, and Halon substitutes. (**NOTE:** Refer to Chapter 1 for additional discussion on Halon and Halon substitutes.) The areas in which these agents are most commonly used include the following:

- Radio and television equipment rooms
- Computer rooms
- Areas containing equipment of high value
- Flammable liquid storage and dispensing areas
- Printing presses

This section covers the so-called *clean* agents and their various properties. It is important to note that the word "clean" is a reference to the amount or type of residue left after the agent's use, not to a health benefit.

Carbon Dioxide Extinguishing Systems

Like most fire extinguishing systems, a carbon dioxide (CO_2) system can be designed for local application and total flooding configurations; both configurations are common. Hand-held hose and standpipe systems are also used, although they are not common and are not covered in this manual. All carbon dioxide systems should be designed, installed, and maintained in accordance with the standards set forth in NFPA 12, *Standard on Carbon Dioxide Extinguishing Systems* 2002 edition. Carbon dioxide is considered a "clean" agent; however, its health ramifications are discussed later in this section. Each gaseous extinguishing agent extinguishes fire in a different way. Unlike water, carbon dioxide has little cooling effect on a fire; it extinguishes fire by displacing air. Carbon dioxide gas settles over the hazard, displaces the surrounding air, and smothers the fire by eliminating oxygen (**Figure 7.9**). Although some ice crystals will accompany the carbon dioxide, the agent has little cooling effect on the fire in these applications.

The most serious problem involving carbon dioxide systems, especially total flooding systems, is personnel safety. Eliminating oxygen from a fire also eliminates it from the surrounding air. Total flooding systems are designed to deliver at least a

Figure 7.9 Although carbon dioxide is a clean agent, it is hazardous because it displaces oxygen.

34 percent concentration of carbon dioxide into an enclosed area. This means that once the system has activated, carbon dioxide will make up at least 34 percent of the atmosphere. Concentrations this high are lethal and people have been killed from these systems. For this reason, total flooding systems must be provided with predischarge alarms as well as discharge alarms. **NOTE:** Pure oxygen is present in only 21 percent of the air we breathe. The bare minimum humans need for life is around 16 percent. When a carbon dioxide extinguishing system is activated in a confined area, the available oxygen is near zero.

A predischarge alarm notifies those present that the system is about to activate. There is a time delay before the system actually discharges the agent. However, alarms alone are not enough to ensure the safety of personnel. All affected personnel must be educated about the dangers of carbon dioxide (**Figure 7.10**). They must also be trained in proper emergency procedures relative to system discharge. Failure of alarms to operate or inadequate training could result in injuries or death.

If a carbon dioxide total flooding system has discharged, no one should enter the area without using self-contained breathing apparatus (SCBA)

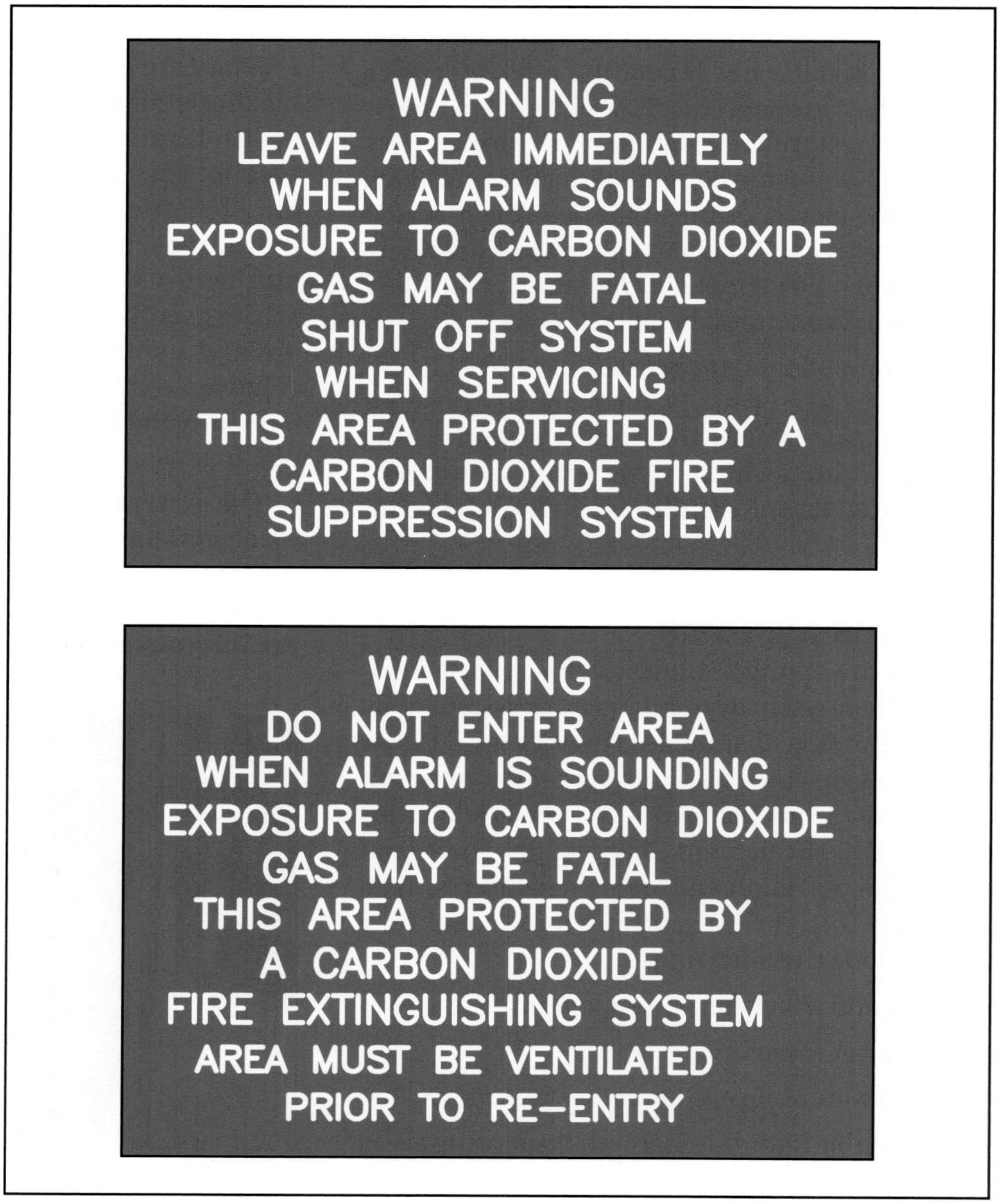

Figure 7.10 Warning signs used within and adjacent to a space protected by a carbon dioxide (CO_2) extinguishing system. It is imperative that employees know the dangers of carbon dioxide and what to do if the system discharges.

WARNING
LEAVE AREA IMMEDIATELY
WHEN ALARM SOUNDS
EXPOSURE TO CARBON DIOXIDE
GAS MAY BE FATAL
SHUT OFF SYSTEM
WHEN SERVICING
THIS AREA PROTECTED BY A
CARBON DIOXIDE FIRE
SUPPRESSION SYSTEM

WARNING
DO NOT ENTER AREA
WHEN ALARM IS SOUNDING
EXPOSURE TO CARBON DIOXIDE
GAS MAY BE FATAL
THIS AREA PROTECTED BY
A CARBON DIOXIDE
FIRE EXTINGUISHING SYSTEM
AREA MUST BE VENTILATED
PRIOR TO RE-ENTRY

Figure 7.11 A local application system will deliver agent directly onto the hazard.

Figure 7.12 Normal manual operation should enable individuals to trigger the extinguishing system as they exit the area.

or until the area has been declared safe. If the local fire department responds to the situation, fire brigade members should brief them about the hazards and proper actions required for entry.

Local application systems are not as dangerous as total flooding systems. In local application systems, carbon dioxide is delivered directly onto the hazard as opposed to filling an enclosure with gas **(Figure 7.11)**. Not much danger exists if the local application system is located outdoors or in a large building.

System Components

The components of the carbon dioxide system are similar to other special extinguishing systems. These components include actuation devices, agent storage containers, piping, and nozzles. The following are three means of actuation for carbon dioxide systems:

- Automatic operation — Operation that is triggered by a product-of-combustion detector.

- Normal manual operation — Operation that is triggered by a person manually operating a control device and putting the system through its complete cycle of operation, including pre-discharge alarms **(Figure 7.12)**.

- Emergency manual operation —This operation causes the system to discharge immediately and without any advance warning to personnel in the area **(Figure 7.13)**.

Figure 7.13 If the emergency warrants, extinguishing agent can be discharged without warning to personnel.

Historically, fixed-temperature or rate-of-rise detection has been used for automatic actuation. However, modern systems may use smoke detectors or even flame-detection equipment to activate carbon dioxide systems. All of these actuation methods trigger control valves on the carbon dioxide supply that allow agent to enter the system and discharge.

Carbon dioxide systems exist as either high-pressure systems or as low-pressure systems. In the high-pressure system, the carbon dioxide is stored in standard DOT cylinders at a pressure of about 850 psi (5 950 kPa) (**Figure 7.14**). The low-pressure system is for much larger hazards. The carbon dioxide in these systems is stored in large, refrigerated tanks where the liquefied carbon dioxide is stored at a temperature of 0°F (-18°C) (**Figure 7.15**). The carbon dioxide is stored at about 300 psi (2 100 kPa) at that temperature.

In either a low-pressure or a high-pressure system, the containers are connected to the discharge nozzles through a system of fixed piping. The requirements for piping are similar to those used in other gaseous systems. Nozzles for total flooding systems may be of the high- or low-velocity type. However, high-velocity nozzles promote better disbursement of the agent throughout the entire area. Local application nozzles are typically of the low-velocity type. This reduces the possibility of splashing the burning material.

Inspection and Testing

Because of the relative complexity of carbon dioxide systems, routine inspection and testing should only be performed by trained personnel. Maintenance and testing should be performed annually by competent contractors, preferably licensed representatives of the system manufacturer. Fire service personnel and facility maintenance personnel can inspect for physical damage to components, check for corrosion or rust, and observe for a change in hazard. Agent cylinders should be inspected semiannually or on a schedule that reflects their use, and changed if necessary. It is also important that an on-site representative read the service tag to determine if or when service should be scheduled.

Figure 7.14 High-pressure carbon dioxide systems.

Figure 7.15 Low-pressure carbon dioxide may be stored in large, refrigerated tanks. *Courtesy of Chemetron Fire Systems.*

Foam Extinguishing Systems

A foam extinguishing system is used where water alone may not be an effective extinguishing agent. As with other types of fixed fire suppression systems, the primary purpose of a fixed foam system is to control and/or suppress a fire in its early stages, thereby reducing or eliminating the need to employ manual methods of fire fighting. Foam systems can also be activated to suppress fire vapors before a fire begins. The type of system and foam to be used will depend upon the hazard that is being protected.

Typical locations for foam extinguishing systems include flammable liquids processing or storage facilities, aircraft hangars, and rolled pa-

Figure 7.16 Rolled paper storage facilities are frequently protected by foam systems.

per or fabric storage facilities (**Figure 7.16**). Before entering into a discussion on foam systems, it is necessary to look at how foam extinguishes a fire, how it is generated, and the various types of foam concentrates that are available.

NOTE: For more detailed explanations of foam equipment and foam delivery systems, consult the IFSTA **Principles of Foam Fire Fighting** manual.

System Description

A foam system must have an adequate water supply, a supply of foam concentrate, piping system, proportioning equipment, and foam makers (discharge devices). There are five types of foam extinguishing systems:

- Fixed
- Semifixed Type A
- Semifixed Type B

- High-Expansion
- Foam/Water

Each of these systems is discussed in the following section.

NOTE: Standards and requirements for foam system design, placement, and other technical information can be found in NFPA 11, *Standard for Low- Medium- and High-Expansion Foam Systems,* 2002 edition; NFPA 13, *Standard for the Installation of Sprinkler Systems;* NFPA 16, *Standard for the Installation of Foam-Water Sprinkler and Foam-Water Spray Systems,* 2003 edition; NFPA 30, *Flammable and Combustible Liquids Code*; and NFPA 409, *Standard on Aircraft Hangars,* 2004 edition. Directions in the design and requirements of foam sprinkler systems can also be found in publications from the manufacturer, Underwriters Laboratories, and FM Global.

Fixed Foam System

A fixed foam extinguishing system is a complete installation that is piped from a central foam proportioning station. A fixed system automatically applies foam to the target hazard (**Figures 7.17 a and b, p. 228**). The foam is discharged through fixed delivery outlets to the hazard being protected. If a pump is required, it is usually permanently installed. A fixed system may be the total flooding type or the local application type that uses either low-, medium-, or high-expansion foam. Most fixed systems are of the deluge type and require actuation by some sort of product-of-combustion detection system. Deluge systems have an unlimited water supply and may also have large foam supplies.

Semifixed Type A System

In a semifixed Type A system, the foam discharge piping is in place but is not attached to a permanent source of foam (**Figure 7.18, p. 229**). The semifixed Type A system requires a separate source for foam solution (usually the fire brigade or fire department). This type of system is found in settings that involve several similar hazards. This system is used primarily on flammable liquid storage tanks. The foam source is a mobile foam solution apparatus (usually a fire truck). This system can be compared to dry standpipe systems.

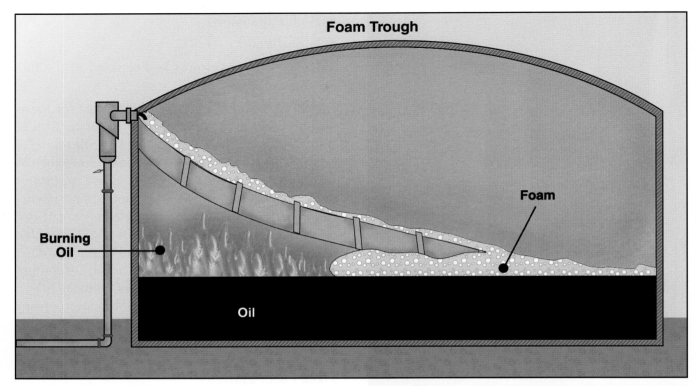

Figure 7.17 a In this fixed system the foam trough gently applies foam to the surface of the fuel.

Many subsurface injection systems are of this type. Subsurface injection is used in situations where topside application may not be effective because of wind or heavy fire conditions. Because the foam solution is lighter than the product in the tank, it will float to the top when introduced at the bottom of the tank. This is a highly effective method for extinguishing bulk tank fires.

Semifixed Type B System

A semifixed Type B system provides a foam solution source that is piped throughout a facility, much like a water distribution system. The foam solution is delivered to foam hydrants for connection to hoselines and portable foam application devices. The difference between a semifixed Type B system and a fixed system is that a fixed system actually applies foam to the hazard while the semifixed system merely provides foam capability to an area. From this point, the foam must then be applied manually.

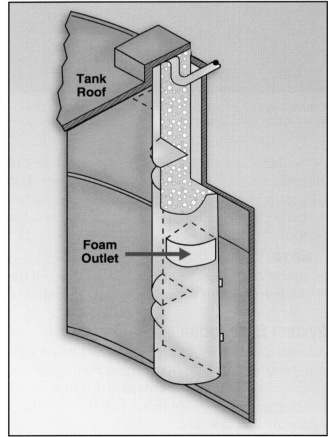

Figure 7.17 b A cutaway view of a foam side dump.

Figure 7.18 The semifixed Type A system is basically a dry standpipe system used to protect flammable liquid storage.

High-Expansion Foam System

A high-expansion foam system is designed for local application or total flooding in commercial and industrial applications. It consists of the following primary components:

- Automatic detection or manual actuation system
- Foam generator
- Piping from the water supply and foam concentrate storage tank to the generator

The high-expansion foam system can be actuated by any of the common fire detection devices or by a manual pull station. The foam generators are powered by electric or gasoline motors or by water. The generators should have a fresh-air intake to make sure that foam does not become contaminated by products of combustion. In addition, venting should be provided ahead of the foam to allow it to move through the area to be protected. In the total flooding application, entire buildings can be filled to several feet (meters) above the highest storage area or equipment within a few minutes.

Foam/Water System

A foam/water system is similar to a deluge sprinkler system but has foam designed into the system (**Figure 7.19**). A foam/water system is used where there is a limited foam concentrate supply but an unlimited supply of water. Thus, if the foam concentrate supply becomes depleted, the system will continue to operate as a straight deluge sprinkler system.

Typically, the foam/water system is an automatic system that operates in the same manner as a regular deluge system. The major differences are that the foam induction system and special aerating sprinklers are at the end of the piping. This type of system produces a lean foam solution that is expanded six to eight times when it is discharged from the sprinkler. This produces a very fluid foam that will flow around obstructions after it is delivered.

The entire foam/water system may be divided into two parts: the water system and the foam system. The foam system contains a concentrate tank, pump, metering valve, strainer, piping, and

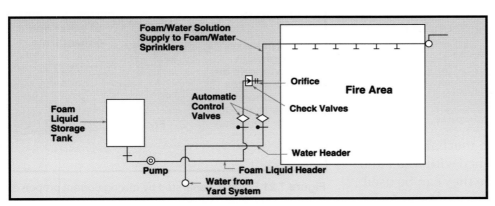

Figure 7.19 The components of a foam/water sprinkler system.

the actuation unit. The water system components are the same as those described in Chapter 6 for deluge sprinklers.

Protein, fluoroprotein, and aqueous film forming foam (AFFF) concentrates may be used in these systems. AFFF may be discharged through regular water sprinklers with favorable results. When discharged through standard sprinklers, AFFF has greater velocity than when it is discharged through foam sprinklers. This tends to improve the spray and the penetration.

The system operates when the initiation devices sense the presence of fire and send an appropriate signal to the system control unit. This in turn triggers the deluge valve and the system. At the same time, the deluge valve on the foam side of the system opens, the foam pump starts, and the concentrate is introduced into the water flow. The foam/water solution flows to the sprinklers where air is introduced to make the foam. The foam is then delivered to the target area.

After the fire is out, the system must be shut down, drained, and thoroughly flushed to remove any foam residue. The concentrate tanks must then be refilled; once this is complete, the valves can be reset and the system restored to service.

How Foam Extinguishes Fire

In general, foam works by forming a blanket on the burning fuel. The foam blanket excludes oxygen, stops the burning process, and cools adjoining hot surfaces. The following methods describe how foam extinguishes fire (**Figure 7.20**).

- Smothering/suppressing — Prevents air and flammable vapors from combining.

- Separating — Intervenes between the fuel and the fire.

- Cooling — Lowers the temperature of the fuel and adjacent surfaces.

- Suppressing — Prevents the release of flammable vapors.

How Foam Is Generated

Foams in use today are of the mechanical type; that is, they must be proportioned with water and aerated (mixed with air) before they can be used. Before discussing types of foams and the foam-

making process, it is important to understand the following terms (**Figure 7.21**):

- **Foam concentrate** — The raw foam liquid prior to the introduction of water and air. Foam concentrate is usually shipped in 5-gallon (20 L) pails or 55-gallon (220 L) drums. Foam concentrate for fixed systems is stored in large fixed tanks that hold 500 gallons (2 000 L) or more.

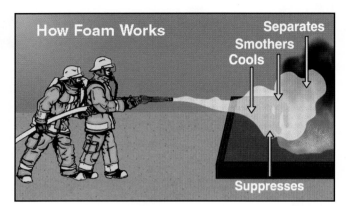

Figure 7.20 Foam works by a variety of methods.

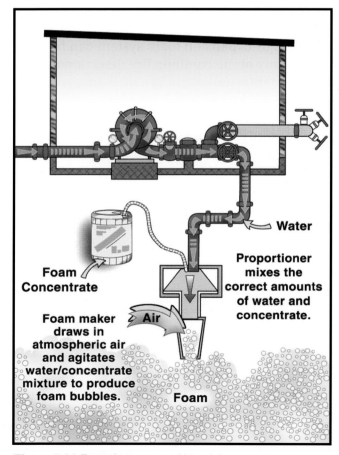

Figure 7.21 Foam is generated by mixing correct proportions of water, foam concentrate, and air.

- **Foam proportioner** — The device that introduces the correct amount of foam concentrate into the water stream to make the foam solution.

- **Foam solution** — A homogeneous mixture of foam concentrate and water prior to the introduction of air.

- **Foam** — Once air is introduced into the foam solution, the completed product is called foam (also known as finished foam).

Four elements are necessary to produce high-quality fire fighting foam: foam concentrate, water, air, and mechanical agitation **(Figure 7.22)**. All of these elements must be present and blended in the correct ratios. Removing any element will result in either no foam or an unusable liquid.

There are two stages in the formation of foam. First, water is mixed with foam liquid concentrate to form a foam solution. This is known as the proportioning stage of foam production. Second, the foam solution passes through the piping or hoseline to a foam nozzle or sprinkler and the foam nozzle or sprinkler aerates the foam solution to form finished foam.

Proportioning equipment and foam nozzles or sprinklers are engineered to work together. Using a foam proportioner that is not hydraulically matched to the foam nozzle or sprinkler, even if the two are made by the same manufacturer, can result in unsatisfactory foam or no foam at all. There are numerous appliances for making and applying foam. A number of these are discussed later in this chapter.

Foam Proportioners

The correct proportioning of foam concentrate into the fire stream requires equipment that operates within strict design specifications. In general, foam proportioning devices operate on one of two basic principles:

- The pressure of the water stream flowing through an orifice creates a venturi action that inducts (drafts) foam concentrate into the water stream.

- Pressurized proportioning devices inject foam concentrate into the water stream at a desired ratio and at a higher pressure than that of the water.

There are several types of foam proportioners in common use for fixed systems. These include line eductors, balanced pressure proportioners, around-the-pump proportioners, pressure proportioning tanks, coupled water motor-pump proportioners, and premixed systems. This section details the various types of foam proportioning devices commonly found in fixed extinguishing systems.

Line Eductors

The line eductor is the simplest and least expensive proportioning device. It has no moving parts in the waterway, which makes it durable and dependable. The line eductor may be attached to the hoseline or may be part of the nozzle **(Figure 7.23, p. 232)**. The two types of line eductors are the in-line eductor and the self-educting master stream nozzle.

NOTE: Line eductors are not used on fixed systems.

Both types of eductors use the *venturi* principle to draft foam concentrate into the water stream. As water at high pressure passes over a reduced opening, it creates a low-pressure area near the outlet side of the eductor. This low-pressure area creates a suction or draft effect. The eductor pickup tube is

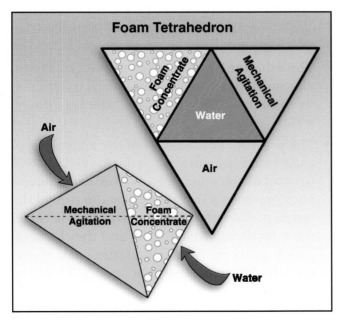

Figure 7.22 As with fire, there are four elements necessary to produce foam.

Figure 7.23 This line eductor is attached to the hoseline.

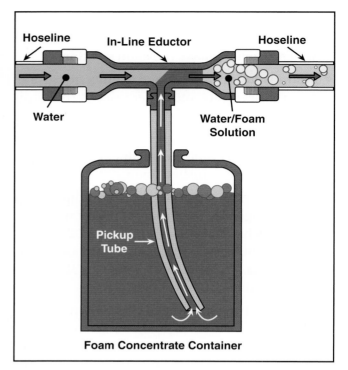

Figure 7.24 The venturi principle. As water under pressure passes over the opening, foam concentrate is drawn into the tubing, creating foam solution.

connected to the eductor at this low-pressure point **(Figure 7.24)**. The pickup tube submerged in the foam concentrate draws concentrate into the water stream, creating a foam water solution.

The in-line eductor is the most common type of portable foam proportioner used in the fire service. This eductor is designed to be attached directly to the pump panel discharge or connected in the middle of a hose lay. When using an in-line eductor, it is very important to follow the manufacturer's instructions about inlet pressure, maximum hose lay from the eductor, and the appropriate nozzle.

In-line foam eductors should not be higher than 6 feet (2 m) above the foam concentrate container from which they are drawing. Lifting foam concentrate more than 6 feet (2 m) may cause proportioning inaccuracy or a total concentrate shutdown. Pickup tubes, however, can be longer than 6 feet (2 m); if the pickup tube needs to be lengthened, 10

feet (3.1 m) can be added. The reason manufacturers supply a short tube is because of concentrate lift restrictions. If longer tubing is required, increase the inside diameter of the tube.

Eductors are relatively tolerant of different operating conditions, but there are several very important operating guidelines that must be observed to obtain high-quality foam:

• The eductor must control the flow of foam concentrate through the system. If it does not, foam concentrate will not be inducted into the water.

• The pressure at the outlet of the eductor (also called back pressure) must not exceed 65 to 70 percent of the eductor inlet pressure. Eductor back pressure is determined by the sum of the nozzle pressure, friction loss in the hose between the eductor and the nozzle, and the elevation (head) pressure. If back pressure is excessive, no foam concentrate will be inducted into the water.

NOTE: Back pressure is not to be confused with nozzle reaction pressure.

- Foam solution concentration is only correct at the rated inlet pressure of the eductor, usually 150 psi to 200 psi (1 050 kPa to 1 400 kPa). Using eductor inlet pressures lower than the rated pressure for the eductor will result in rich foam concentrations. Conversely, using inlet pressures greater than the rated inlet pressure will produce lean foam concentrations. Neither a too rich nor too lean concentration will work properly.

- Eductors must be properly maintained and flushed after each use. Flush the eductor by submerging the foam pickup tube in a pail of clean water and inducting water through it for at least one minute **(Figure 7.25)**. Check the strainer after each use and clean it if necessary.

- Set metering valves to match the foam concentrate percentage and the burning fuel. Failure to do so will result in poor quality foam.

- The foam concentrate inlet to the eductor should not be more than 6 feet (2 m) above the liquid surface of the foam concentrate. If the inlet is too high, the foam concentration will be very lean or foam may not be inducted at all.

In order for the nozzle and eductor to operate properly, both should have the same rating in gpm (L/min). It is important to remember that the eductor, not the nozzle, must control the flow. If the nozzle has a lower flow rating than the eductor, the eductor will not be able to pick up concentrate. An example of this situation would be a 60 gpm (240 L/min) nozzle with a 95 gpm (380 L/min) eductor.

Figure 7.25 Following each use, thoroughly flush foam eductors, nozzles, and hoselines to prevent corrosion of metal parts.

Using a nozzle with a higher rating than the eductor also gives poor results. A 125 gpm (500 L/min) nozzle used with a 95 gpm (380 L/min) eductor will result in improper eduction of the foam concentrate. Low nozzle inlet pressure, however, will result in poor-quality foam because of the lack of proper aeration.

Self-educting handline foam nozzles were once common but are rarely used today. Their primary disadvantage was their lack of mobility; they had to be operated next to the foam concentrate containers. Self-educting master stream foam nozzles, on the other hand, are still commonly used. Because they are master stream devices, mobility at the fire scene is not critical.

Balanced Pressure Proportioner

Using a balanced pressure proportioner is one of the most reliable and accurate methods for proportioning foam **(Figure 7.26, p. 234)**. The primary advantage of the balanced pressure proportioner is its ability to monitor the demand for foam concentrate and to adjust the amount of concentrate being supplied. Another major advantage of a balanced pressure proportioner is its ability to discharge foam from some outlets and plain water from others at the same time. Thus, a single fire pump can supply both foam attack lines and protective cooling water lines simultaneously. Supplying different types of lines is not possible with around-the-pump proportioners.

Systems equipped with a balanced pressure proportioner have a foam concentrate line connected to each fire pump discharge outlet or the system riser **(Figure 7.27, p. 234)**. This line is supplied by a foam concentrate pump that is separate from the main fire pump. The foam concentrate pump draws the concentrate from a supply source (fixed tank and/or auxiliary supply). This pump is designed to supply foam concentrate to the outlet at the same pressure at which the fire pump is supplying water to the discharge. A hydraulic pressure control valve jointly monitors pump discharge and foam concentrate pressure from the foam concentrate pump. This control valve ensures that the concentrate pressure and water pressure are balanced.

Figure 7.26 A balanced pressure proportioner.

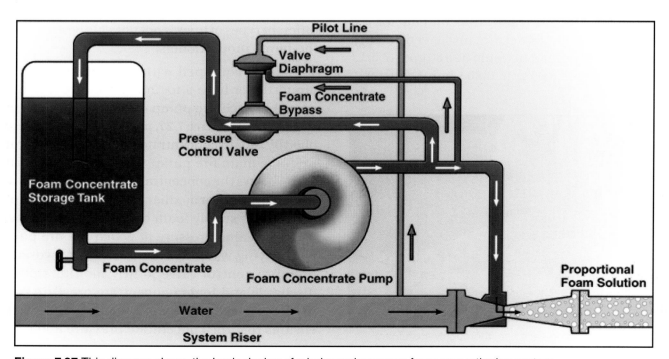

Figure 7.27 This diagram shows the basic design of a balanced pressure foam proportioning system.

The orifice of the foam concentrate line is adjustable at the point where it connects to the riser. If 3 percent foam is used, the foam concentrate discharge orifice should be set to 3 percent of the total size of the water discharge outlet. If 6 percent foam is used, the foam concentrate discharge orifice is set to 6 percent of the total size of the riser, and so on. Because the foam and water are being supplied at the same pressure and the sizes of the discharges are proportional, the foam is proportioned correctly.

Around-the-Pump Proportioner

The around-the-pump proportioner is especially useful when the water pressure is low or when a motor is not available for a separate foam concentrate pump. This type of proportioner is the most common type of built-in proportioner installed in mobile fire apparatus today. It is also installed on some fixed systems.

This system runs a small return line from the discharge side of the pump back to the intake side of the pump **(Figure 7.28)**. An in-line eductor is positioned on the pump bypass. This unit is rated for a specific flow and should be used at this rate, although it does have some flexibility. For example, a unit designed to flow 500 gpm (2 000 L/min) at a 6 percent concentration will flow 1,000 gpm (4 000 L/min) at a 3 percent rate.

A major disadvantage of the around-the-pump proportioner is that the pump cannot take advantage of incoming pressure. If the inlet water supply is any greater than 10 psi (70 kPa), the foam concentrate will not be able to enter into the pump intake. There are versions of these systems that are capable of handling intake pressures of up to 40 psi (280 kPa). Another disadvantage is that the pump must be dedicated solely to foam operation. An around-the-pump proportioner does not allow plain water and foam to be discharged from the pump at the same time.

Portable around-the-pump foam proportioning kits have recently become available. They are sometimes called "the portable crash truck." These kits attach to the auxiliary suction valve of the apparatus. They are driven by a short hose that is connected from the pump's discharge back into the pump inlet via the around-the-pump unit.

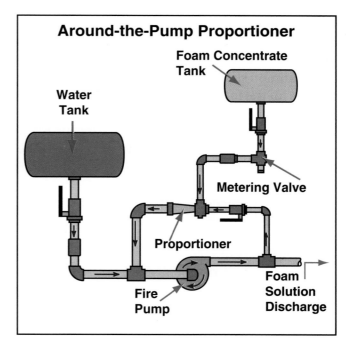

Figure 7.28 Around-the-pump proportioners are commonly found on structural fire apparatus.

These kits are very reliable and require little or no maintenance. They offer flows from 50 to 2,000 gpm (200 L/min to 8 000 L/min) of foam solution. Like a portable foam eductor, this kit uses no electronic interface, flowmeter, or foam concentrate pump. Smaller versions are also available for use with Class A foam or 1 percent AFFF.

Pressure Proportioning Tank System

The pressure proportioning tank system consists of one or two foam concentrate tanks that connect both to the water supply and to foam solution lines of the overall system. This system is designed so that a small amount of water from the supply source is pumped into the concentrate tank(s). This water volumetrically displaces the concentrate into the foam solution line where it is mixed with discharge water. A bladder membrane inside the tank separates the water pumped into the tank from the foam concentrate expelled. The system allows for automatic proportioning over a wide range of flows and pressures, and it does not depend on an external power source. The system is limited by the size of the concentrate tank. Once the concentrate is expended, the water must be drained from the tank before it can be refilled with concentrate.

Coupled Water Motor-Pump Proportioner

This proportioner consists of two positive-displacement rotary-gear pumps that are mounted on a common shaft. One pump is for water and the other is for the foam concentrate. The water pump is proportionally larger than the foam pump. As water flows through the larger pump, it causes the smaller pump to turn and draft foam concentrate from the foam tank. Because the pumps are sized in proportion to each other, the correct foam/water solution is produced. This type of proportioner is used in fixed-system applications. It is limited to only two sizes, and both are designed for 6 percent proportioning rates. One type flows 60 to 180 gpm (240 L/min to 720 L/min) and the other flows 200 to 1,000 gpm (800 L/min to 4 000 L/min).

Premixed Foam

The most common application for premixed foam is in fire extinguishing equipment that uses stored pressure for discharge energy. Examples are hand and wheeled portable fire extinguishers, skid-mounted twin-agent units, and rapid intervention vehicle-mounted systems (**Figures 7.29 a and b**). With this equipment, the foam solution is stored in a pressure vessel charged with nitrogen. The agent is discharged when the nozzle is opened. The main disadvantage with this equipment, as well as the previously mentioned discharge method, is that the supply of agent is limited and the entire unit must be recharged to continue application.

Foam Nozzles and Sprinklers

Foam nozzles and sprinklers, sometimes referred to as foam makers, are the devices that deliver the foam to the fire or spill. Fixed systems may have both handlines and foam sprinklers or generators attached to them. Standard fixed-flow or automatic water fog nozzles may be used on AFFF or FFFP handlines. IFSTA defines a handline nozzle as "any nozzle that one to three firefighters can safely handle that flows less than 350 gpm (1 400 L/min)." Most handline foam nozzles flow considerably less than this amount. However, aerating low-expansion foam nozzles will produce higher quality finished foam than will water fog nozzles. The various types of foam nozzles and sprinklers are listed as follows:

Figures 7.29 a and b Premixed foam units are available in a number of sizes. The entire unit must be refilled when the solution is used. *Top photo courtesy of ConocoPhillips.*

- Smoothbore nozzle
- Low-expansion foam nozzle
- Self-educting foam nozzle
- Standard fixed-flow fog nozzle
- Automatic nozzle
- Foam/water sprinkler
- High back-pressure foam aspirator
- High-expansion foam generator

Smoothbore Nozzle

The use of smoothbore nozzles is limited to Class A, compressed-air foam system applications. In these applications, the smoothbore nozzle provides an effective fire stream that has maximum reach capabilities (**Figure 7.30**). Tests indicate that the reach of the compressed-air foam system fire stream can be more than twice the reach of a low-energy fire stream.

When using a smoothbore nozzle with a compressed-air foam system, disregard the standard rule of thumb that the discharge orifice of the nozzle be no greater than one-half the diameter of the hose. Tests show that a 1½-inch (38 mm) hoseline may be equipped with a nozzle tip up to 1¼-inch (29 mm) in diameter and still provide an effective fire stream.

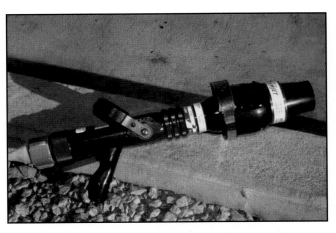

Figure 7.31 A smaller foam nozzle used for generating low-expansion foam.

Figure 7.30 The smoothbore nozzle provides a great reach for foam application.

Low-Expansion Foam Nozzle

Foam solution must be mixed with air to form foam. The most effective appliance for generating low-expansion foam is the air-aspirating foam nozzle. The special design of a foam nozzle aerates the foam solution to provide the highest quality foam possible. Smaller foam nozzles (30 to 50 gpm [120 L/min to 200 L/min]) may be hand held (**Figure 7.31**). Larger foam nozzles (250 gpm [1 000 L/min] and up) may be monitor-mounted units (**Figure 7.32**).

The fog/foam-type nozzle is marketed with two stream-shaping foam adapters. The basic adapter breaks the foam solution into small streams and

Figure 7.32 A monitor-mounted foam unit.

at the same time inducts air through a venturi action. The cone-shaped attachment gives the nozzle extra reach, and the screen produces a more homogenous high-air-content foam for gentle applications. Again, it is necessary to emphasize the importance of using a compatible foam maker and eductor.

Self-Educting Foam Nozzle

The self-educting foam nozzle operates on the same principle as the in-line eductor. The major difference is that this eductor is built into the nozzle rather than into the hoseline (**Figure 7.33**). As a result, its use requires the foam concentrate to be available where the nozzle is operated.

The self-educting master stream foam nozzle is used where flows in excess of 350 gpm (1 400 L/min) are desirable (**Figure 7.34**). Lack of mobility is not a problem with self-educting master stream foam nozzles because master streams are stationary. This eductor also requires a specific inlet pressure for proper operation. This inlet pressure tends to be somewhat lower than that of in-line eductors.

The self-educting foam nozzle has several advantages:

- It is easy to use.
- It is inexpensive.
- It works with lower pressures than required by in-line eductors.
- A wide variety of flow rate versions are available.

There are several disadvantages to using a self-educting foam nozzle. First of all, if the foam nozzle is moved, the foam concentrate will also need to be moved. The logistical problems of relocation are magnified by the gallons of concentrate that must also be moved. Use of a foam nozzle eductor also compromises firefighter safety; firefighters cannot move quickly and they must leave foam concentrate behind if they are required to back out for any reason.

Standard Fixed-Flow Fog Nozzle

The fixed-flow, variable-pattern fog nozzle is used with foam solution to produce a low-expansion, short-lasting foam. This nozzle breaks the foam solution into tiny droplets and uses the agitation of water droplets moving through air to achieve its foaming action. Its best application is when it is used with regular AFFF concentrate and Class A foams because its filming characteristic does not require a high-quality foam to be effective. These nozzles cannot be used with protein and fluoroprotein foams or any alcohol-type foams. The fixed-flow fog nozzle may be used with alcohol-resistant AFFF foams on hydrocarbon fires but should not be used on polar solvent fires. This is because the aeration produced by this nozzle is not sufficient to handle polar solvent fires. Some nozzle manufacturers have foam aeration attachments that can be added to the end of the nozzle to increase aeration of the foam solution. Water fog nozzles have found a growing acceptance because of their ability to be used as both a standard fire fighting nozzle and as a foam nozzle (**Figure 7.35**).

Figure 7.33 Portable self-educting foam nozzles, such as this unit intended for wildland fire fighting, must have the foam concentrate located near the nozzle.

Figure 7.34 A master stream appliance applying foam in the rain-down method.

Figure 7.35 Water fog nozzles are effective for producing low-expansion short-lasting foam.

Automatic Nozzle

An automatic nozzle operates with an eductor in the same way that a fixed-flow fog nozzle operates. However, the eductor must be operated at the inlet pressure for which it was designed, and the nozzle must be fully open. An automatic nozzle may cause problems if the eductor is operated at a lower pressure than that recommended by the manufacturer or if the nozzle is gated down.

Foam/Water Sprinkler

Foam/water sprinklers are found on fixed-foam deluge and foam/water systems (**Figure 7.36**). These systems use AFFF or FFFP foams. Many foam/water sprinklers resemble aspirating nozzles in that they use a venturi velocity to mix air into the foam solution. Some systems use standard sprinklers to form a less-expanded foam through simple turbulence of the water droplets falling through the air. Foam/water sprinklers come in upright and pendant designs (**Figures 7.37 a and b, p. 240**). Their deflectors must be adapted to meet the specific installation requirements.

High Back-Pressure Foam Aspirator

High back-pressure foam aspirators, or forcing foam makers, are in-line aspirators used to deliver foam under pressure. These foam makers are most commonly used in subsurface injection systems

Figure 7.36 Foam/water sprinklers are used to protect electric transformers like the one shown. *Courtesy of Kidde Fire Fighting.*

that protect cone-roof hydrocarbon storage tanks, but they are also used in other applications. High back-pressure aspirators supply air directly to the foam solution through a venturi action (**Figure 7.38, p. 240**). This action typically produces a low-air-content foam that is homogenous and stable.

High-Expansion Foam Generator

High-expansion foam generators produce high-air-content, semistable foam. The air content ranges from 200 parts air to one part foam solution (200:1) to 1,000 parts air to one part foam solution (1,000:1). There are two basic types of high-expan-

Figures 7.37 a and b Upright and pendant sprinkler installations.

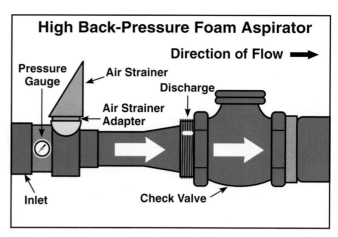

Figure 7.38 Because of the high back pressure created by the fuel in the storage tank, high back-pressure foam aspirators may be installed in the subsurface foam injection line. Air pressure is added to the foam to help it overcome the resistance of the fuel.

Figure 7.39 Medium- and high-expansion foam handlines use special nozzles to discharge foam.

sion foam generators: the mechanical blower and the water-aspirating type. The water-aspirating type is very similar to the other foam-producing nozzles except that it is much larger and longer. The back of the nozzle is open to allow airflow (**Figure 7.39**). The foam solution is pumped through the nozzle in a fine spray that mixes with air to form a moderate-expansion foam. The end of the nozzle

has a screen or series of screens that breaks up the foam and further mixes it with air. These nozzles typically produce a lower-air-volume foam than do mechanical blower generators.

A mechanical blower generator is similar to a smoke ejector in appearance (**Figure 7.40**). It operates on the same principle as the water-aspirating nozzle except that the air is forced through the foam spray instead of being pulled through by the water movement. This device produces a higher-air-content foam and is typically associated with total flooding applications.

Figure 7.40 Mechanical foam blowers are commonly used for high-expansion foams in total flooding systems.

Foam System Inspection and Testing

As with many of the other types of special extinguishing systems covered in this chapter, foam systems are highly complex and require specially trained personnel to inspect, service, and test them for readiness. All valves and alarms attached to the system should be checked semiannually. Foam concentrates, foam equipment, and foam proportioning systems should be checked annually (**Figure 7.41**). Qualitative tests should be performed on the concentrate to ensure that no contamination is present. The concentrate tank should be checked for signs of sludge or deterioration.

Figure 7.41 This portable foam unit is a multiagent system that includes premixed foam and dry chemical extinguishing agents.

Summary

Special extinguishing systems are needed when application of water is not the best way to extinguish a fire, either because of the nature of the hazard itself or because large amounts of water will ruin specialized equipment. In these situations, specialized, controlled application of wet or dry chemicals, gaseous agents, or foam are better choices. Because these systems are often custom designed and all contain only a specific amount of extinguishing agent, they must be installed carefully and inspected regularly to ensure successful operation. Many of the chemicals or gaseous agents used in specialized systems are dangerous to health, so it is imperative that people who may be affected be made aware of the potential hazards. Individuals must know how to exit an area promptly and activate the extinguishing system from a position of safety. Even carbon dioxide, which is present in the air we breathe, is hazardous if it is present in too-high a concentration.

Because each special extinguishing system is designed and installed to protect a particular hazard or commodity, it is very important that fire personnel be aware of the agent used in the system, its suitability, its operating readiness, and methods for inspecting each system.

Appendix A
NFPA Job Performance Requirements (JPRs) with Page References

NFPA 1001 JPR Numbers	Fire Detection and Suppression Systems Page References
5.3.10	90 – 92
5.3.16	7 – 12, 23 – 27, 39 – 47
6.5.1	51 – 56, 66 – 76, 104 – 117, 166 – 194

NFPA 1002 JPR Numbers	Fire Detection and Suppression Systems Page References
5.2.1	92 – 117
5.2.2	92 – 117
5.2.4	117 – 119

NFPA 1031 JPR Numbers	Fire Detection and Suppression Systems Page References
4.3.16	92 – 117
4.3.5	104 – 119, 123 – 148, 151 – 160, 194 --210
4.3.6	76 – 84
4.3.7	7, 12, 32 – 39
5.4.3	24 – 31, 36 – 38
5.3.4	24 – 31, 51 – 52, 56 – 84, 123 – 148, 159 – 160, 194 – 210
5.3.8	166 – 173
5.3.8	7, 15 – 23, 28 – 29.54 – 56, 66 – 76, 222 – 230
5.4.3	56 – 84, 123 – 139, 151, 194 – 210.
6.3.4	24 – 31, 56 – 63, 64 – 76
6.3.8	11 – 14, 15 – 23, 66 – 76
6.3.12	76 – 84
7.3.1	165 – 173, 173 – 194
7.3.5	153
7.3.6	104 – 117
7.3.10	24 – 31, 56 – 76, 104 – 119, 123 – 148, 151 – 159, 194 – 210, 219 – 241
8.3.5	11 – 31, 76 – 81, 123 – 148

NOTE: For more information about extinguisher tests, consult Underwriters Laboratories UL 711, *Rating and Fire Testing of Fire Extinguishers.*

Class A Rating Tests
Wood Crib Test

The wood crib test is conducted by using a wood crib built of dried spruce or fir lumber having a moisture content between 9 and 13 percent. The nominal dimensions of the members are 2 × 2 inches. For higher rating tests, the dimensions are 2 × 4 inches. The wood crib is elevated to the specified height according to the desired rating for which it is being tested. The ignition of the crib is accomplished by neutral n-heptane that is placed in a pan beneath the crib **(Figure B.1)**. The amount of n-heptane required for ignition of the crib is determined by the desired rating. The preburn is also an essential variable of the rating system. The Class 1-A to 4-A rating is to have a preburn period of 8 to 10 minutes or until the stick members in the top three rows have been reduced to diameters of ¾ to 1 inch (19 mm to 25 mm). An extinguisher that can extinguish a fire of this size will receive a 1-A rating. A 2-A extinguisher will extinguish a

Figure B.1 The wood crib is ignited. *Courtesy of Tyco Safety Products.*

fire that is twice as large, and so on. Extinguishers tested for ratings above 4-A are required to have a preburn period of 6 to 7 minutes or until stick members in the top three rows have been reduced to diameters of one- half to two-thirds of the original dimension.

The initial attack is started from 6 feet (2 m) away and is directed at the front of the crib **(Figure B.2)**. The duration of the extinguisher's discharge should be 13 seconds or greater for extinguishers with a rating of 2-A or greater.

NOTE: For extinguishers rated 10-A or higher, only the wood crib test is used for rating purposes. The following tests are omitted in this case.

Wood Panel Test

This test is conducted by using a solid-wood, square-panel backing. Two horizontal sections of furring strips, spaced apart and away from the panel by vertical furring strips, are applied to the backing. The panel is constructed of spruce or fir lumber, with a moisture content between 9 and 13 percent, that is kiln dried **(Figure B.3)**. The backing panel itself has nominal lumber dimensions of 1 x 6 inches, while the furring strips consist of ¾-× ¾-inch fir. Four excelsior windrows made of seasoned basswood, poplar or aspen are placed at the base of the test panel. Before ignition, the wood panel itself is sprinkled with No. 2 fuel oil. Between 4½ and 5½ minutes after ignition, the furring strips will fall away from the panel. Five seconds after the furring strips fall away, the fire attack is started from a distance of 10 feet (3.05 m) from the face of the test panel.

Class B Rating Tests

The test is conducted using a pan of the proper size for the extinguisher desired, for example, a 2-inch (50 mm) layer of n-heptane in an 8-inch (200 mm) deep pan. The area of the fuel and fire must correspond to 1 square foot (0.09 m^2), and so on. Upon ignition, the n-heptane is allowed to burn for 60 seconds. The attack is then started using an appropriate Class B extinguisher (dry chemical, carbon dioxide etc.) **(Figure B.4)**.

Figure B.2 The attack is begun at the proper distance: 6 feet (2 m). *Courtesy of Tyco Safety Products.*

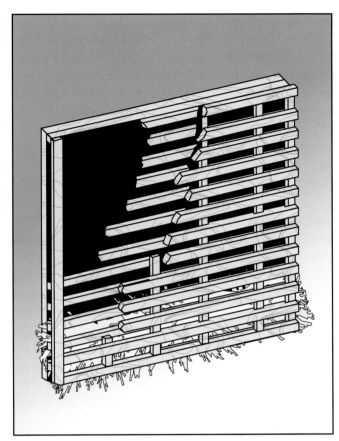

Figure B.3 Excelsior is placed at the bottom of the wood panel.

During the Class B extinguisher fire tests, the following factors are evaluated:

- Time of extinguisher application
- Time at which the fire was extinguished or controlled
- Time the agent discharge ended

Class C Rating Tests

The extinguisher is mounted on an insulated platform and discharged at a target made of two 12 × 12-inch (300 mm by 300 mm) copper sheets. The distance between the extinguisher nozzle and the target is approximately 10 inches (250 mm). The electrical potential between the target and the nozzle is 100,000 volts. The extinguisher is discharged at the target once for 20 seconds. The extinguisher is discharged a second time for 15 seconds. To pass the test, there are to be no visible effects of the agent discharge against the target.

NOTE: The Class C designation is given to extinguishers only in conjunction with other previously established ratings, such as Class A, Class B, etc.

Figure B.4 Extinguishing a Class B pan fire.

Class D Rating Tests

Class D fire extinguishers or manually applied extinguishing agents must be capable of extinguishing fires in designated materials and prevent burning material from scattering as the agent is applied. Tests are conducted for magnesium chips and for materials made of sodium and potassium.

Magnesium

Several tests are conducted in magnesium chips:

- Area fire test
- Pallet transfer fire test
- Premix fire test
- Casting fire test

Area Fire Test

The area fire test simulates a fire occurring in magnesium chips and dust in either a dry state or when mixed with cutting oils. The test makes use of two grades of magnesium chips. When tested in an oily state, the chips and dust are mixed with a petroleum-based cutting oil to 10 percent by weight. The chips and dust for the test fires are spread evenly on a dry steel plate.

The fuel is ignited by directing a gas and air flame onto the surface of the fuel bed at the center. Application of the extinguisher is begun when either the fire has spread over 50 percent of the fuel surface or when a deep-seated fire is observed within the fuel. The extinguisher discharge is begun at a distance of no more than 8 feet (4.4 mm) from the edge of the test fire.

Pallet Transfer Fire Test

This test simulates a fire in magnesium chips on a combustible surface. The pallet transfer test uses two wooden pallets. On one pallet a pile of chips is placed. The other pallet is covered with a layer of the extinguishing agent 1 inch (25 mm) deep. The pile of chips on the first pallet is ignited. When the fire involves 50 to 75 percent of the fuel, it is transferred to the second pallet by two technicians using shovels. The fire is then extinguished by application of additional agent.

Premix Fire Test

This test checks the ability of an extinguishing agent to inhibit the spread of a fire through a pile of chips that has been mixed with the agent. The test is conducted using three 10-pound (4.5 kg) piles of magnesium chips that are mixed with different amounts of the extinguishing agent. The chips are arranged in a conical pile on a steel plate and ignited. The fire is then attacked with additional agent.

Casting Fire Test

In the casting fire test, a magnesium casting of approximately 25 pounds (11.3 kg) is placed on a dry steel plate and ignited. A fire involving a casting presents special difficulties because it has both horizontal and vertical surfaces. After ignition, the fire is allowed to burn until a pool of molten metal has developed. The fire is then attacked with the extinguishing agent being applied, either by manual means or from an extinguisher. The method of application depends on the agent and the manufacturer's recommendations.

Sodium, Potassium, and Sodium-Potassium Alloy Tests

To test extinguishing agents for these types of fires, two tests are employed:

- **Spill fire test.** In this test, 3 pounds (1.36 kg) of metal are melted to ignition. The burning fuel is then poured onto a steel pan having an area of 4 square feet ($0.37m^2$) and a depth of 6 inches (150 mm). The fire is then attacked either manually or from an extinguisher with the extinguishing agent.

- **Pan fire test.** The pan fire test consists of two separate fires. Two pans are used that are 2½ square feet ($0.23m^2$) and 6 inches (150 mm) in depth. In one pan, a 7-pound (3.2 kg) charge of combustible metal is used, resulting in molten metal at a depth of about 1 inch (25 mm). In the other pan, 35 pounds (15.9 kg) of combustible metal is used, resulting in a pool of molten metal approximately 3½ inches (90 mm) deep. This deeper pool of burning fuel may prove more difficult to control, and special techniques may be necessary.

Class K Rating Tests

Fryer Test Method

The test is to be conducted using a commercial deep fat fryer and vegetable cooking oil. The purpose of the test is to demonstrate the extinguisher's ability to do the following:

- Extinguish the fire in the fryer
- Prevent reignition of the vegetable oil for 20 minutes or until the temperature of the oil decreases to a set temperature
- Cause no splashing of flaming oil outside the fryer

Appendix C
Pump Tests

There are two categories of pump tests that must be performed on a stationary fire pump. The first is the initial or acceptance test, which is performed by the pump manufacturer, the engine manufacturer, the controller manufacturer or other factory-authorized representatives. Much the same as testing a fire suppression pump mounted on a pumping apparatus, the acceptance test is one conducted prior to the ownership of the pump being transferred to the purchasing end user.

Prior to the acceptance test, all local authorities should be notified of the testing time and location and invited to participate. During the test, all peripheral equipment that affects the operational capabilities of the pump is to be included in the test. The multiple control interwiring, the pressure maintenance pump, and any alternate power supply systems are to be completed, checked, and approved by the electrical contractor before testing begins.

The second category of stationary fire pump tests includes those of a periodic nature: weekly, monthly, quarterly, semi-annually, or annually. Usually associated with normal maintenance inspections, these tests are intended to ensure the continued acceptable performance of the fire pump and its associated systems. Periodic testing and maintenance is normally the responsibility of the owner/operator of the facility where the pump is installed. Occasionally, through maintenance agreements, insurance requirements, or other pump maintenance and performance contracts, the original manufacturer may retain responsibility for the continued performance of the pump. It is incumbent upon the local fire authority to clarify these issues before granting approval of and for the fire pump installation.

Acceptance Tests

The primary performance criteria for standard fire pumps are contained in NFPA 20, *Standard for Installation of Stationary Pumps for Fire Protection,* 2003 Edition. Additionally, Factory Mutual Research Corporation provides Pump Acceptance Test Data check sheets that also contain information on industrial fire pumps. The material contained in these checklists is compiled from the requirements of NFPA 20, which allows the inspector to systematically check the components and performance of the fire pump. Prior to the installation of the pump and before any acceptance testing, the manufacturer must provide a certified shop (pump) test characteristic curves showing head capacity and the brake horsepower of the fire pump. This is used to compare the expected performance with the fire pump's actual performance during the test. Once installed, a new industrial fire pump must be capable of satisfying three test points.

- The first point for the vertical-shaft pump is the maximum point of 140 percent of rated pressure at shutoff.

- The second point is a minimum of 100 percent of rated capacity at 100 percent of rated pressure.

- The third point specifies 65 percent of rated pressure while delivering 150 percent of the rated number of gpm.

NOTE: Another component of a field acceptance test includes the flushing and hydrostatic testing of all suction and discharge piping associated with the operation of the fire pump. The installing contractor is required to furnish a certificate indicating that flushing and the hydrostatic testing have been completed. Many local authorities issue permits and successful test certificates after they have witnessed these procedures. Check with the local authority having jurisdiction for their requirements.

NOTE: For more information about the specifics of acceptance testing, consult the FPP *Fire Protection and Water Supply Analysis* manual.

Periodic or Routine Fire Pump Maintenance and Testing

All fire pumps must be inspected, tested, and have routine maintenance performed on a regular basis as described in NFPA 25, *Standard for the Inspection, Testing and Maintenance of Water-Based Fire Protection Systems*, 2002 Edition. Although maintenance and testing are performed as often as the authority having jurisdiction requires, the standard recognizes that a minimum interval for each of these must be adopted. Weekly tests are more cursory than annual tests and usually involve the activation of the fire pump system while recording pressure gauge information. Annual tests are far more comprehensive. These tests require actual fire flow and the determination of continued satisfactory performance of the fire pump and its associated systems during a calibrated testing format. Every test must be recorded and retained in the permanent inspection and maintenance files for the fire pump.

Weekly Tests

Weekly inspection should address four main areas of concern:

- General housekeeping conditions
- The condition of the pump and the water supply system
- The electrical system
- The power system (driver) that drives the pump

The general condition of the pump room is inspected primarily to ensure adequate heating of the pump room space to prevent the pump and piping from freezing. Additionally, the proper operation of ventilation louvers must be verified so that air for combustion is in abundant supply as well as air for cooling of the motor and fire pump.

All pump system components must be visually inspected for obvious signs of malfunction or damage including pipe/seal leaks, both water and oil. System pressure gauges must be within system operational parameters. The oil in any priming or suction reservoirs must be checked and replenished if it is below manufacturer's recommendations.

The electrical systems must also be in full operating condition. The pilot light on the controller indicating that power is on must be lit as well as that of the transfer switch. The isolating switch is to be closed with the emergency power standby source ready and available. The electrical phase alarm pilot light is to be off or the normal phase rotation light illuminated (three-phase electrical motors can be seriously damaged or destroyed if an "out of phase" electrical current is supplied). Additionally, if one leg of the current is absent or of lower-than-expected current the motor will overheat and burn up.

For diesel engine driver systems, the fuel tank level must be at least two-thirds full. The controller switch is to be in the AUTOMATIC position. All batteries are to be fully charged and all current gauge readings within normal ranges. No alarm pilots may be active. If cooling water is used the reservoir is to be full with no fluid leaks.

Weekly pump tests are also required by NFPA 25. These tests are conducted by automatically starting the fire pump but not by flowing water. Electric pumps are run for 10 minutes and diesel pumps for a minimum of 30 minutes. These tests are to be conducted in the presence of qualified personnel who can make adjustments if any deficiencies are noted. The required performance checks include the pump system where pressure gauge readings are taken and the pump starting pressure is recorded. Pump packings are checked for leakage and adjusted if necessary. Unusual vibrations and noises during the operation are checked and adjustments are made to eliminate them. The packing boxes, bearings, and pump casing are monitored for overheating.

Electrical system procedures include observing and recording the time it takes the recorder to accelerate to full speed. Record the time the controller is on the first step (for reduced voltage or reduced current starting).

When testing a diesel engine system, check and record the time for the engine to crank and the time it takes for the engine to reach its running speed. The engine oil pressure gauge and the speed, water, and oil temperature indicators are to be monitored while the engine is running.

A steam system is to have the steam pressure checked and recorded as well as the observed time for the turbine to reach running speed.

Annual Tests

Annual fire pump assembly tests are to be conducted to ensure the performance of the pump at minimum, rated, and peak flows. These tests are performed by flowing and measuring the pump suction, discharge pressure, and each hose stream supplied by the pump. The pressure relief valve must be monitored to ensure that the design pressure of the system is not exceeded and is observed during each flow condition (no-flow, flow) to determine if it operates at the proper pressure.

For fire pumps that feature an automatic transfer switch, several additional tests are necessary to ensure that overcurrent protective devices do not open. To conduct this test, a simulated power failure condition occurs when the pump reaches peak load. The transfer switch should operate, shifting to the alternate power source. A successful test will be recorded once it is verified that the pump continues to operate at peak load. Once this has been successfully completed, the power failure condition is removed. After the designated time delay the pump is reconnected to the normal power supply system.

It is not acceptable to increase the engine speed past the rated speed of the pump to reach rated pump performance. The results of the tests must be evaluated by qualified individuals. The pump assembly will be given an acceptable test performance evaluation if the test matches the initial unadjusted field acceptance test curve of the pump or if the fire pump matches the performance characteristics listed on the pump nameplate.

NOTE: For an in-depth description of the weekly and annual testing procedures for a fire pump, refer to NFPA 25, *Inspection, Testing, and Maintenance of Water-Based Fire Protection Systems*.

Appendix D
Fire Department Operations With Standpipes

Standpipes are critical to effective fire fighting in high-rise buildings. Fire departments responding to buildings that are equipped with standpipes must understand how the systems are designed and how they operate. Emergency responders should conduct pre-incident surveys to identify the following:

- Location of fire pump rooms
- Size and method of operation of fire pumps
- Number of zones serving high-rise buildings
- Location of fire department connections (interior and exterior)

Fire departments responding to buildings with standpipes should develop standard operating procedures (SOPs). NFPA 13E, 2000 edition, *Guide for Fire Department Operations in Properties Protected by Sprinkler and Standpipe Systems*, will be of assistance in preparing adequate procedures. It is imperative that fire departments understand how standpipes operate *prior* to conducting emergency operations at the site. Pre-incident planning is necessary to determine the true flow capability of the system.

The construction of high-rise buildings is such that a fire on the upper floors may not be apparent to companies arriving at the building entrance. Therefore, tactical operations must proceed on the assumption that a fire exists until the fire floor has been reached and the nature of the incident has been determined. The arriving units should perform the following:

- Accurately identify the fire floor.
- Establish control of the lobby.
- Evaluate conditions on the fire floor.
- Designate a forward command post and staging area two floors below the fire floor.

- Provide for life safety of persons in immediate danger.
- Ensure an adequate water supply.
- Begin an initial attack on the fire.
- Check for fire extension and occupants on floors above the fire.

In buildings equipped with standpipes, fire department operations begin at the base of the standpipe where one of the first-arriving engine companies attaches water supply lines to the fire department connection **(Figure D.1)**. This proce-

Figure D.1 A firefighter is connecting the pumper to the fire department connection.

dure is used to charge a dry system or to augment the pressure in a wet system. To adequately supply the fire fighting lines using the standpipe, the pump operators or officers should be familiar with the location of fire department connections and the nearest fire hydrant.

Pump discharge pressure depends on the following elements:

- Friction loss in the hose deployment from the pumping apparatus to the fire department connection

- Friction loss in the hose on the fire floor

- Nozzle pressure for the type of nozzle employed

- Elevation head pressure loss due to the height of the fire floor

- Friction loss in the standpipe

Generally, the friction loss in the standpipe is small unless the flows are large, such as when two 2½-inch (65 mm) lines are being supplied from the same riser. Allowances should be made for pressure losses in the fire department connection and pipe bends.

When a standpipe system is known to be equipped with pressure-reducing valves, the elevation pressure used must be based on the *total height* of the standpipe or the zone being used. This is because the pressure-reducing valve generally acts to reduce whatever pressure is presented to it. If the pressure at a certain pressure-reducing valve is less than that for which the valve was adjusted, the result will be inadequate pressure for the hoselines. For additional information regarding hydraulic calculations for these circumstances, refer to the IFSTA **Pumping Apparatus Driver/Operator Handbook**.

Another problem occurs when vandals or curious individuals open the hose valves in dry standpipes and leave them in an open position. When the standpipe is charged, water will discharge on levels below the fire floor, and it will be necessary for a firefighter to go down the stairwell to close the valves.

A well-developed standard operating procedure (SOP) that clearly describes standpipe operations must be developed and implemented by the fire

department. As firefighters advance into a standpipe-equipped building, they should take all necessary equipment with them. Many fire departments carry prearranged standpipe packs for such use (**Figure D.2**). Equipment that is necessary for these operations and should be listed in the SOP and includes the following:

- Hose, 150 feet (50 m)

- Valve wheel (in case one is not attached to the valve)

- Nozzle

- Pipe wrench

- Gated wye, 2½ × 1½ × 1½ inches (65 mm by 38 mm by 38 mm)

- Forcible entry tools

- Spanner wrench

- Rope hose tool or other hose securing devices

The need for some of this equipment may become apparent only with experience. For example, valve wheels are sometimes stolen. Valves may also have several years of paint accumulation, making them difficult to operate. Stairwell doors are sometimes locked from the stairwell side, necessitating forcible entry onto the fire floor.

Some fire departments use wheeled handcarts to carry equipment to standpipe connections near the fire area. It may sometimes be convenient to

Standpipe Hose Pack

Figure D.2 A standpipe hose pack.

place the tools in a canvas bag for carrying on the shoulder. Hand lights and portable radios are also necessary on the fire floor. Before firefighters leave the lobby, they should review the building floor plan, if available, and the type of standpipe. Extra hose and air cylinders should be brought into the building and kept in the lobby until needed **(Figure D.3)**. Fire suppression in upper stories utilizing standpipe systems is extremely labor intensive. It is not uncommon to find that several companies of firefighters must be deployed to support one company actually involved with fire combat. Additional personnel may be needed to carry these items to the fire or staging floor. Personnel should be assigned to check riser valves and fire pump operation.

Firefighters should proceed to a location two floors below the fire floor where a forward command post and staging area is established. Before advancing to the fire floor, full personal protective equipment (PPE) must be donned and operational. If elevators are necessary to reach the upper floors of tall buildings, they should be placed in the Firefighters Service Phase II operational mode (Refer to IFSTA **Fireground Support Operations** Manual and ASME/ANSI A17.1 — 2000 Safety Code for Elevators and Escalators). When possible, it is safest to walk up the stairs. Firefighters should connect their standpipe fire hose to the standpipe discharge connection on the floor below the fire and advance the hose to the fire floor **(Figure D.4)**. This practice provides a safe working area in the event of heavy involvement on the fire floor. (**NOTE:** One firefighter must remain at the standpipe valve to charge the line after it has been moved to the entrance of the fire floor.) Excess hose can be laid up the stairway past the fire floor. Gravity will allow the hose to feed more easily down to the fire floor as it is advanced. Care must be exercised to create as little obstruction as possible for occupants exiting the building via the stairwell.

When heavy fire has developed on the fire floor, the line should be charged and the stairwell door carefully opened to begin the attack. If needed, additional hoselines can be attached to the hose valves on lower floors and advanced to the fire floor. The firefighter operating the standpipe valve should watch for fire developing behind the attack team and warn them before their position becomes untenable. Additional hoselines must be deployed

Figure D.3 When entering a burning high-rise building, firefighters should carry extra equipment to Staging.

Figure D.4 Organize the fire attack from the floor below the fire.

to "hold" the stairway and maintain a safe passageway for the buildings occupants.

Firefighters must decide in advance what type of nozzle to use. Fog nozzles produce fine water droplets that afford better protection to advancing firefighters and more efficient heat absorption. However, it is possible to obstruct the fog nozzle if scale is present in the standpipe. Although a solid stream does not produce a spray, it provides good reach and penetration into burning debris.

When a building is equipped with multiple standpipes, more than one riser can be used for a multiple line attack. Firefighters advancing onto a floor from different directions must keep in contact by radio so that they can coordinate their efforts and not endanger each other with opposing fire streams. When dry standpipes are charged, there is often a delay before the delivery of water to the hose valve because of the amount of air that must be expelled from the system. Air can also be trapped and become compressed so that when a valve is opened, the air under high pressure is released. All attack hoselines must be charged and air bled from them before opening the stairway door onto the fire floor.

A number of simple operations can be used by the fire department to overcome various standpipe impairments. For example, if a fire department connection has a frozen swivel, a double male adapter can be used with a double female adapter to duplicate the original connector (**Figure D.5**).

If an exterior fire department connection is totally unusable because of vandalism, the standpipe riser can be charged at the first-floor level by attaching a double female adapter to a hose valve at the first floor level. If an individual hose valve on an upper floor is inoperative, a valve on the next floor down can be used. To overcome a problem in a single-riser building where the standpipe is totally unserviceable, it may be necessary to hoist a line up the outside of the building. Often this can be accomplished by unrolling hose from every second or third floor down the side of the building rather than attempting to hoist line several floors. Standpipes in adjacent buildings can also be used to protect exposures. As a last resort, supply hose can be laid up the interior stairwell to take the place of the standpipe (**Figure D.6**).

If the floor is reachable by aerial apparatus, standpipe connections at the tip of the aerial apparatus can be used to supply hoselines on upper

Figure D.6 Lay the hoseline against the outside wall to avoid sharp bends.

Figure D.5 Adapters connect two threaded couplings of the same thread type.

levels **(Figure D.7)**. If the fire floor is higher than the reach of the aerial device, the aerial device can still be used in place of hoselines for the first 6 to 10 floors. The aerial device will have much less friction loss than would hose used for the same distance. It is important to remember that when an aerial device is committed to standpipe operations, its use as an emergency egress is limited.

Figure D.7 Some elevating platforms have outlets for attaching handlines.

Appendix E
Fire Department Operations at Sprinklered Occupancies

Firefighters' knowledge of sprinkler system operation enhances fire fighting operations in buildings protected by sprinkler systems. At the very least, sprinklers can save fire fighting units a great deal of time and effort. Proper fire fighting tactics in sprinklered properties should proceed in support of an operating sprinkler system. Fire departments should adopt standard operating procedures for pre-incident planning and operations at sprinklered occupancies. Refer to the following NFPA standards for additional information about sprinklered occupancies and dwellings equipped with residential sprinklers:

- NFPA 13, *Standard for the Installation of Sprinkler Systems*
- NFPA 13D, *Standard for the Installation of Sprinkler Systems in One- and Two-Family Dwellings and Mobile Homes*
- NFPA 13R, *Standard for the Installation of Sprinkler Systems in Residential Occupancies up to and Including Four Stories in Height*

Pre-Incident Planning at Sprinklered Occupancies

Fire department standard operating procedures cannot be established unless fire department personnel are familiar with the sprinklered properties in their area of jurisdiction. Inspections and pre-incident planning visits are very useful in developing information necessary to develop standard operating procedures (SOPs). Information that should be noted about the sprinkler system includes the following:

- Buildings and nature of occupancies protected by automatic sprinklers

- Type of sprinkler system(s)
- Water supply to the sprinklers, including available volume and pressure
- Locations of sprinkler valves (interior and exterior) and the area each valve controls
- Locations of fire department connections, the specific area each serves, and the nearest water supply that will not interfere with the supply to the sprinkler system(s) and the location of the nearest fire hydrant **(Figure E.1)**.
- Pumper having primary responsibility for charging the fire department connection
- Alternate means for supplying water to the system in case of damage to the fire department connection, such as the test header at a fire pump

NOTE: It will probably be necessary to open the test header control valve in the pump room.

Figure E.1 Firefighters must locate fire department connections during pre-incident planning so valuable time will not be lost during an emergency.

- Location of spare or replacement sprinklers
- Name of building owner or official to contact in an emergency
- A list of sprinkler system deficiencies
- Location of all sprinkler system risers and the location of fire alarm control units needed to silence and reset the fire alarm system
- Design criteria as shown on the hydraulic nameplate and whether or not the system is a pipe schedule system

Fireground Operation Guidelines

Although individual fire department standard operating procedures may vary, one procedure should be the same for all departments: Upon arrival at a sprinklered property, immediate preparations should be made to supply the fire department connection (FDC). The first-in engine company should locate the FDC and the nearest suitable fire hydrant. If there is any indication of an actual fire, such as smoke showing or the ringing of a sprinkler alarm, a minimum of two 2½-inch (65 mm) hoselines or one 3-inch (77 mm) hoseline should be connected to the FDC (**Figure E.2**). Additionally, the engine company should lay a supply line(s) to the hydrant and make all the appropriate connections.

NOTE: Some fire departments use the second-in engine to connect to the FDC.

The interior attack crews should locate the fire and determine whether charging of the sprinkler system is necessary. It may be argued that it is prudent to charge the system immediately upon arrival of the fire companies. Obviously, this action is appropriate if a fire is evident. In some situations, however, the sprinklers may have extinguished the fire, or the system activation may be a malfunction. In most cases, it is desirable to confirm the presence of fire before pumping into the system (keeping in mind that water damage usually causes more damage than the fire itself).

Should the system need to be charged, the pump operator should slowly develop 150 psi (1 050 kPa) at the engine and maintain this pressure. In many situations, 150 psi (1 050 kPa) may be much more than is needed; however, it is a standard rule of thumb for supply at fire department connections. From outside the building, it is usually difficult to determine how many sprinklers are operating in the fire area and the amount of friction loss in the system as a result. If it becomes obvious that the fire is spreading, additional lines should be connected to the fire department connection and charged. A rapid size-up may be hindered by low visibility because of the smoke being cooled by the sprinklers. This cooling causes the smoke to lose its normal buoyancy.

A firefighter familiar with the building should immediately check the control valves, if they are accessible, to ensure that they are open (**Figure E.3**). Closed valves should be opened except when

Figure E.2 This pumper is supplying water to a fire department connection, which can be used to boost the pressure in the standpipe system.

Figure E.3 The butterfly valve has an indicator that shows the position of the valve.

it is known that the building or area has been undergoing construction or renovation affecting the sprinklers. Opening valves under these circumstances would cause a severe loss of water to the system. The firefighter assigned to this task should carry a flashlight and a portable radio. If a pump supplies the system, the firefighter should also ensure that the pump is running. Control valves are frequently located in the pump room so that both functions can be performed by the same firefighter.

Firefighters should advance handlines as necessary **(Figure E.4)**. A properly designed sprinkler system will have controlled the fire, and only a small-diameter line should be necessary to complete extinguishment. However, there should be no hesitation in advancing larger or additional lines if the fire escalates. IFSTA recommends that handlines for interior structural fire fighting be not less than 1½-inch (38 mm) in diameter and preferably 1¾-inch (44.5 mm).

Through pre-incident planning, fire department personnel should have identified water supplies that will not rob the sprinkler or standpipe system of water. They should make sure that these water sources are the ones used when the actual incident is in progress. The design requirements for hydraulically calculated systems include an allowance for hose streams that are in addition to the sprinkler system demand:

- Light hazard = 100 gpm
- Ordinary hazard = 250 gpm
- Extra hazard = 500 gpm
- High-piled storage = 500 gpm

Therefore, these hose stream flow rates may be used without concern for depleting the water supply to the sprinkler system.

Truck companies should perform ventilation and salvage. The effectiveness of roof ventilation may be reduced by the cooling effect of the sprinkler discharge. In these cases, horizontal ventilation using smoke ejectors or blowers may be more effective **(Figure E.5)**. In the long run, performing horizontal ventilation requires less work and personnel than would be needed to open a roof. This allows more firefighters to perform salvage operations.

Figure E.4 Firefighters in proper position to advance hoselines.

Figure E.5 Blowers are sometimes set up in tandem.

There are situations where the combined efforts of sprinklers and fire companies are inadequate and a major fire develops. In these situations, partial or total structural collapse may occur. This collapse may result in broken sprinkler lines, resulting in a tremendous waste of water through broken lines. If the building is equipped with an outside post indicator valve, this valve(s) can be closed to conserve water.

WARNING!
When fighting fires in a sprinklered building, DO NOT shut the sprinkler system off until the fire has been extinguished.

Many disastrous losses have occurred from shutting off sprinkler systems when the fire "appeared" to be under control. If the area involved in the fire is served by a sectional or floor control valve, that valve, rather than the main riser valve, should be closed to minimize damage to the area affected.

Occasionally during overhaul, buried fire is uncovered and redevelops. For this reason, it is good practice for a firefighter with a portable radio to remain at the control valve until overhaul is completed (**Figure E.6**).

The pump operator should also be ordered to shut down lines to the fire department connection. The piping from the fire department connection may bypass the main valve. In the absence of a sectional valve, water will continue to flow until the pumper is shut down.

Fire crews must give high priority to salvage operations during the fire fighting operations and after the fire has been extinguished. In addition to the products of combustion, a large quantity of water is introduced to the area when sprinklers activate. All contents in the area of operating sprinklers should be covered with salvage covers. Particular attention should be paid to the floor below the one on which the sprinklers are operating. Water may seep down several levels. All contents should be covered, and any objects on the floor that may be damaged by water should be raised.

Figure E.6 Someone should be equipped with a two-way radio and ready to re-open the valve if necessary.

The incident commander must be aware of the weight of the water discharged from the sprinkler system and hose streams, particularly where commodities have a high absorption capability such as in the storage areas of clothing. The added weight could result in structural collapse of the building (**Figure E.7**). For example, where a system is discharging 400 gpm (1 600 L/min) and 300 gpm (1 200 L/min) is absorbed by the commodity and 100 gpm is runoff, approximately 2,500 pounds (1 136 kg) is added to the weight of the building PER MINUTE. In ten minutes, this is 25,000 pounds (11 360 kg). For more information on salvage operations, see the IFSTA **Loss Control** manual.

Until the officer in charge has determined that a fire is sufficiently extinguished, the control valve must not be closed. In order to decrease water damage, sprinkler *stops* can be used to shut off the flow of water at the operating sprinklers. Firefighters can carry several types of sprinkler tongs and wedges in the pockets of their turnout coats. Other types of tongs are attached to poles, which eliminates the need for a ladder. A wedge is made of soft wood and is designed to be driven into position

Figure E.7 To reduce collapse potential and to help reduce fire loss, water should be removed as soon as possible.

Figure E.8 To stop the water, insert small wooden wedges between the discharge orifice and the detector.

with the heel of the hand until it is wedged between the sprinkler frame and the orifice (**Figure E.8**). If the wedge is properly made, practically all of the water flow can be stopped. Sprinkler tongs are more effective in stopping the water flow because the rubber or neoprene seal used eliminates dripping when applied properly.

Fireground Operations in One- and Two-Family Dwellings Equipped with Residential Fire Sprinklers

Residential sprinkler systems are designed to provide a higher level of life safety than is provided by smoke detectors and building materials in one- and two- family dwellings. They are less complex than commercial fire sprinkler systems, yet have unique features that must be taken into consideration when working in dwellings where they are installed.

Most fire departments have their own operating procedures for residential structure fires. The same procedures that are normally followed by a department during suppression activities in non-sprinklered dwellings should be followed in a fire involving a residence with fire sprinklers. The main differences between a commercial occupancy having fire sprinklers and a dwelling with a residential fire sprinkler include the following:

- There is no fire department connection to boost the sprinkler system pressure and flow.

- Sprinkler piping is usually CPVC plastic and is not as strong as the piping installed in a commercial fire sprinkler system (**Figure E.9, p. 268**).

- Restoring the system is usually much easier following an activation than restoring a commercial fire sprinkler system.

- Residential fire sprinkler systems are most often a branch water line connected to the residential service pipe.

Residential sprinklers are designed for life safety and fire control, not property conservation. Because dwellings typically have a relatively low fire load, this type of sprinkler system uses much less

Figure E.9 Residential sprinkler piping is usually CPVC plastic.

ditional water needs are calculated in the original size-up and are provided as needed by the engine companies advancing suppression hoselines into the dwelling.

Most residential fire sprinklers, depending upon local water authority requirements, are connected directly to the residential water service line and not directly to a water main. NFPA 13D allows several different configurations for water supply design to be employed with these systems. Local water authority requirements and regulations should be determined and then compared with those configurations outlined in NFPA 13D prior to installation.

NFPA 13D allows a considerable amount of leeway in the integrity of sprinkler piping in residential systems. These systems are designed to be affordable for the average homeowner; therefore, industrial-strength piping is not necessary. This is important to keep in mind during ventilation and overhaul operations. Residential sprinkler piping is often designed for CPVC plastic piping, copper piping, or thin-wall steel piping installation; each is susceptible to damage during ordinary fire fighting operations. While the sprinkler system can be shut down, this option may not be advisable, especially if the fire is not yet fully extinguished. A pre-incident familiarization is recommended for all sprinklered buildings, which would include residential buildings.

System restoration is simply a matter of replacing the fused sprinkler(s) and turning the sprinkler valve back on. Although an alarm device may be installed, it will not be the type that has to be reset in order to place the system back in service.

water than a commercial system. A typical residential fire sprinkler system is designed to deliver from 13 to 23 gpm (52 L/min to 92 L/min) for a period of 10 minutes. There is no fire department connection (FDC) in the front of a private residence and no way to support the sprinkler system directly. Ad-

In order for sprinkler systems to function properly, they must be properly designed, installed, and maintained. Although the fundamental concept of the automatic sprinkler system is simple, ensuring proper design can become very complex because of the variety of buildings and hazards encountered in actual practice. An inadequately designed system cannot be expected to control fires.

Early in the design process, the designer of an automatic sprinkler system must establish several fundamental aspects of the sprinkler system. These fundamentals include the following:

- Available water supply
- Building construction
- Freezing potential
- Occupancy classification
- Commodity classification
- Value of commodities or operation to be protected

This information will determine the most appropriate type of system (wet, dry, deluge, etc.) and the required water supply.

The designer must also determine and specify the type of sprinklers, sprinkler spacing, type of pipe, pipe sizes, pipe hangers and supports, valves, alarms, drains, and other details. Because the design of modern sprinkler systems is complicated, this work is performed by fire protection engineers who work for engineering and contracting firms that specialize in sprinkler system design and installation.

Sprinkler systems are usually designed according to NFPA standards listed earlier in this manual. These standards contain information on the many aspects of sprinkler systems and are almost universally used for system design. Other standards that address special situations include NFPA 409, *Standard on Aircraft Hangars;* NFPA 230, *Standard for the Fire Protection of Storage,* and NFPA 430, *Code for the Storage of Liquid and Solid Oxidizers.*

Occupancy Classifications

The design of a sprinkler system begins with a determination of the occupancy to be protected. NFPA 13 classifies occupancies into four very broad classifications that refer only to the sprinkler system installation and its water supply. The four classifications are as follows:

- **Light-Hazard.** Areas where the quantity and/or combustibility of contents is low. Firefighters can expect a low rate of heat release if these occupancies catch fire. There are no subclasses within the light-hazard classification.

- **Ordinary-Hazard (Groups 1 and 2).** Ordinary-hazard occupancies are subdivided further into two subclasses that are designated ordinary hazard Group 1 or Group 2, depending on the amount and configuration of combustibles.

  — Ordinary-Hazard Group 1 includes occupancies such as bakeries, canneries, and electronic plants where the combustibility of the contents is low, the quantity of combustibles is moderate, the heat release rate is moderate and combustibles do not exceed 8 feet (2.4 m) in height.

  — Ordinary-Hazard Group 2 includes occupancies such as cereal mills, confectionery products manufacturing, printing processes, and tire manufacturing where the quantity and combustibility of the contents is moderate to high, the heat release rate is moderate to high, and combustibles do not exceed 12 feet (4 m) in height.

- **Extra-Hazard (Groups 1 and 2)**. Extra-hazard occupancies are those where severe fires can be expected. These occupancies are further divided into Group 1 and Group 2 categories:

  — Extra-Hazard, Group 1. These occupancies regularly operate with high levels of dust, lint, or flammable liquids, and present the probability of a rapidly developing fire that is accompanied by high rates of heat release.

  — Extra-Hazard, Group 2. These occupancies contain moderate to substantial amounts of combustible liquids or where it is difficult to reach combustibles. Examples would be flammable liquid spraying, plastics processing, solvent cleaning, and varnish and paint dipping.

These classifications are not precise; occupancies that generally fall within one class may have characteristics of another. For example, offices are generally classified as light hazard, but storerooms within an office building may have a higher fuel load than would normally be found within an office. Nonetheless, these classifications serve as a starting point for sprinkler system design. Many requirements for the design of a fire sprinkler system are based on the occupancy classification. Therefore it is extremely important to determine the correct occupancy classification.

NOTE: Refer to Chapter 1 of this manual for additional discussion on occupancy classifications.

Just as it affects systems that are designed using a pipe schedule, building occupancy also affects hydraulically designed systems. Fire development is affected by the amount of fuel, the type of fuel, and the fuel configuration (see discussion in the following section). These factors influence the rate of fire growth and the rate of energy release; that is, the severity of the fire. The sprinkler system is designed to control the potential fire in the various spaces within the structure it protects. Occupancies that present a greater fire hazard require higher discharge densities and areas of application.

Sprinkler Spacing

The spacing of sprinklers is a fundamental aspect of sprinkler system design and is affected by occupancy classification. Sprinklers must be located throughout the area protected in such a manner that their discharge can cover the entire area and control fire. They must also be spaced so that they can respond quickly to the heat of a developing fire.

NFPA 13 limits the maximum protection areas per sprinkler. The typical sprinkler system design uses protection areas less than the maximum permissible areas due to pressure limitations. These spacing limitations are subject to modification depending upon several design factors. The sprinkler contractor or fire protection engineer will determine the proper spacing according to state codes as the design of the system progresses. Other spacing limitations exist, such as distance between branch lines and distance between individual sprinklers on a branch line.

Sprinkler Pipe Sizing

From a design standpoint, next to the sprinklers themselves, probably the most important part of the sprinkler system is the pipe. Because the pipe's function is to route water from the source of supply to the sprinklers, the sprinkler pipe must be of a size to provide an adequate amount of water to the operating sprinkler or sprinklers. Although this factor is a simple concept, it becomes more complex when it is realized that there is no way to know in advance how many sprinklers or which sprinklers may operate when a fire occurs. A 1-inch (25 mm) pipe cannot supply ten sprinklers, yet it is uneconomical to use a 6-inch (150 mm) pipe if only two or three sprinklers are operating. There are two methods for determining pipe size: the pipe schedule method and hydraulic calculations.

Pipe Schedule Method

NFPA 13 contains requirements to size piping according to the number of sprinklers supplied by the pipe. This is called the *pipe schedule*. The differences in pipe size per occupancy classification are based on lower-hazard occupancies requiring less water than higher-hazard occupancies. The number of sprinklers that can be supplied by a given pipe size is based on the occupancy classification. For example, while a light-hazard pipe schedule system will permit a 2½-inch (65 mm) pipe to supply 30 sprinklers, an ordinary hazard pipe schedule system will permit the 2½-inch (65 mm) pipe to supply only 20 sprinklers. Typically, pipe schedule systems are more expensive than hydraulically calculated systems because they require larger cross-main and feed main pipe sizes.

The objective of sprinkler system design is to ensure that enough water is discharged through the sprinklers at the proper pressure to control a fire. While generally acceptable when used with an adequate water supply, the use of the pipe schedule in designing systems is not precise. Unless the design is hydraulically calculated, it is difficult to determine that the sprinklers will discharge the proper flow rate until a fire occurs. This "trial by fire" method is not generally considered satisfactory when the structure being protected is worth millions of dollars. In addition, strict application of the pipe schedule results in large-diameter pipes, 6 or 8 inches (150 mm or 200 mm), which are heavy and costly.

Hydraulic Calculations

It is possible to make use of hydraulics to calculate the pressure losses of water flowing through sprinkler pipe. These calculations ensure that the system will meet minimum design discharge requirements. Computer programs that consider all factors of water supply, piping, valves, fittings, hydraulic characteristics of sprinklers, and the required water discharge density or flow rate of the sprinklers do most hydraulic calculations of a sprinkler system. The great majority of sprinkler systems being installed today make use of hydraulic design.

The discharge density is expressed in gallons per minute per square foot (gpm/ft^2) or liters per minute per square meter ($L/min/m^2$). The sprinkler system is designed to discharge this minimum density over a specified area. For example, a sprinkler system could be designed to discharge 0.20 gpm/ft^2 (8.1 $L/min/m^2$) over an area of 2,000 square feet (186 m^2). This would be a total flow of approximately 400 gpm (1 600 L/min). The actual flow rate

will be higher than that indicated by multiplying 0.20 gpm/ft^2 times the area of sprinkler operation due to pressure balancing in the system.

The hydraulic calculation method assumes the fire will activate only those sprinklers in a certain area, typically 1,500 ft^2 (139.3 m^2). The amount of water that is discharged from those sprinklers is based on the required density and the protection area per sprinkler. The required density is based on the occupancy classification and the area of sprinkler operation. The protection area per sprinkler is regulated by NFPA 13 based on the occupancy classification and the type of construction at the ceiling or roof.

For example, consider a grocery store that is classified as an Ordinary Hazard Group 2. The required density is 0.20 gpm/ft^2 (8.1 mm/min) if the area of sprinkler operation is 1,500 ft^2 (139.3 m^2). Assuming the protection area per sprinkler is 100 ft^2 (93 m^2), the minimum discharge from each sprinkler in the hydraulically most demanding 1,500 ft^2 (139.3 m^2) area is 20 gpm (80 L/min).

The specified discharge density is a "minimum." The actual discharge density is neither uniform nor constant over the design area. The pressure loss due to the flow of water in the sprinkler pipes results in different pressures at each sprinkler. Normally, the pressure increases upstream toward the water supply. As more sprinklers open in the course of a fire, more water flows. The increased flow causes the pressure at individual sprinklers to change, resulting in variation in actual discharge from each successive sprinkler along a branch line.

The use of hydraulic design results in systems tailored to a given occupancy classification. If the occupancy changes from one classification to another, the original system design may be inadequate. Once a hydraulically designed system has been installed, its design density and area of operation cannot be determined by casual examination. Therefore, the adequacy of a system to protect a proposed new occupancy may be in doubt. To overcome this difficulty, durable placards are attached to the system's riser. Most codes require the following minimum information to be imprinted on the placard:

- Location of design area
- Discharge densities of design area
- Required flow and residual pressure demand at the base of the riser
- Hose stream demand
- Occupancy/commodity classification
- Maximum permitted storage height and configuration

For additional information on hydraulic calculations for sprinkler systems, see the FPP *Fire Protection Hydraulics and Water Supply Analysis* manual.

Appendix G
Storage Methods Used in Warehouses

There are four general storage methods that can be used in warehouses. These methods are storage in bulk, solid piling, storage on pallets, and rack storage. Each method has its own fire-related characteristics. The most significant characteristic is the extent of the horizontal and vertical spaces that may exist within the storage configuration.

Bulk Storage

Bulk storage is used for materials such as sugar, flour, and plastic pellets. Most of these materials are stored in bins or silos, but open piles on the floor of a building are sometimes found. One hazard of bulk storage includes possible production of explosive dust during handling. Another hazard is that fire can spread rapidly on conveyors or through conveyor housings beyond the reach of sprinklers. Bulk storage of some materials can result in spontaneous ignition within the pile. Fire control requires soaking and/or digging into the pile with front loaders.

Solid Piling

Solid piling consists of simply stacking the merchandise cartons or cases directly on top of each other for the full height of the stack. Obviously, it is only practical where the containers have enough strength to support several upper tiers. Lift trucks equipped with a side clamp, hand trucks, or conventional forklift trucks do material handling.

Because solid piling contains no horizontal voids, it presents fewer paths for fire development than other storage methods. However, vertical fire spread is still possible up the sides of combustible packaging. Pile stability can also be a problem

Figure 6.1 Rolled paper storage can be hazardous to firefighters. If the bottom rolls become wet, the entire stack can collapse.

when cartons are weakened either by fire or by the soaking action of hose streams or sprinklers **(Figure G.1)**.

Pallet Storage

Palletized storage is widely used because loads can be easily handled with forklift trucks **(Figure G.2, p. 274)**. Loads consisting of individual containers or groups of individual containers (boxes, crates, drums, etc.) are made into roughly the shape of a cube the size of the pallet: 3 to 4 feet (1 m to 1.2 m) on a side. The pallet forms a platform beneath the load that can be lifted and readily moved by a forklift truck. Depending on the structural strength and weight of the loads, pallets can be stacked several layers high.

Pallets can be made of several different materials, wood being the most common. The horizontal channels into which the prongs of the forklift fit create combustible channels. Fire can spread

Figure 6.2 Pallet storage can be more easily handled by forklifts.

Figure 6.3 Rack storage is used for large loads that may or may not be placed on pallets.

rapidly through these spaces because they are shielded from overhead sprinkler discharge. Plastic pallets are gaining in popularity. They are made of high-density polyethelene and can contribute significantly to a fire. Where plastic pallets are used in a warehouse, the sprinkler system should be checked to determine that it is capable of fire control in this area. Idle pallet storage (particularly plastic pallets) may create a greater challenge than the commodities stored on the pallets.

The variety and size of forklift trucks available make palletized storage practical and flexible; however, the method has limitations. Aisles must be wide enough to accommodate the forklift trucks and height is limited by the strength or shape of the packaging. Both of these shortcomings can be overcome and warehouse efficiency can be increased by the use of rack storage.

Rack Storage

Rack storage involves the use of a structural framework to support storage containers or loads on pallets **(Figure G.3)**. Materials-handling equipment consisting of specially designed forklift trucks with high reach or automated stackers is used to store and retrieve the loads. Rack heights of 25 feet (8 m) are typical, although heights of 60 feet (20 m) are not uncommon. Racks can be designed in single or double rows. They can also be designed with rollers to accept a number of palletized loads in depth.

The use of automated stackers further increases operating efficiency, but it complicates fire control. The automated equipment can operate in comparatively narrow aisles, which permits communication of fire from rack to rack. This situation becomes especially difficult when the fire spreads vertically up the front plane of the material being stored.

Rack storage has provided a very serious challenge to fire protection engineers. Within the rack structure, horizontal air spaces exist at each level of storage. Fires burning through these spaces are not only shielded from sprinklers but also may be located far below the ceiling sprinklers. The rack configuration also provides vertical channels for fire spread. These same factors also complicate manual fire fighting.

Installing sprinklers only at the ceiling level is generally inadequate for protection of rack storage. When rack storage exceeds 12 feet (4 m) in height, sprinklers may be required within the racks themselves, depending on the class of commodity being stored and the method of packaging. If rack storage exceeds 25 feet (8 m) in height, sprinklers are required within the racks in all cases. If the rack storage involves the use of solid shelves, sprinklers may be required beneath each shelf in double-row racks regardless of height.

The sprinklers installed within the racks are supported from the rack structure (**Figure G.4**). The actual engineering of in-rack sprinkler protection is complex and involves the determination of discharge density, sprinkler spacing and positioning, and overall water demand. All of these items are dependent on the commodity class, rack height, and specific rack type. NFPA 13 contains the specific details for the proper design of in-rack sprinklers based on the given conditions.

The sprinklers installed within rack storage must be provided with water shields to prevent them from being cooled (and thus not operational) by the discharge from in-rack or ceiling sprinklers above. Sprinklers installed within racks normally have ½-inch (13 mm) orifices, have an ordinary temperature rating, and are either upright or pendant types.

Because of its importance in modern warehousing and the potential for large loss, protection of rack storage has undergone a considerable amount of investigation and experimentation by such organizations as Underwriters Laboratories and Factory Mutual Research Corporation. The provisions of NFPA 13 are a result of some of that research. Research and investigation efforts continue as fire experience and rack storage technology continue to develop.

Figure 6.4 Sprinklers may be installed within racks to provide additional fire protection.

Glossary

A

Acceptance Testing — Preservice tests on fire apparatus or equipment performed at the factory or after delivery to assure the purchaser that the apparatus or equipment meets bid specifications.

Actual Delivered Density (ADD) — is a measure of the amount of water discharged that actually reaches the fuel surface.

AFFF — Abbreviation for aqueous film forming foam.

Alarm-Initiating Device — Mechanical or electrical device that activates an alarm system. There are three basic types of alarm-initiating devices: manual, products-of-combustion detectors, and extinguishing system activation devices.

Alarm Check Valve — Type of check valve installed in the riser of in an automatic sprinkler system that transmits a water flow alarm when the water flow in the system lifts the valve clapper.

Altitude — Geographic position of a location or object in relation to sea level. The location may be either above, below, or at sea level.

Annunciator Panel — Device that indicates the source of fire alarm activation.

Atmospheric Pressure — Force exerted by the atmosphere at the surface of the earth due to the weight of air. Atmospheric pressure at sea level is about 14.7 psi (101 kpa). Atmospheric pressure increases as elevation is decreased below sea level and decreases as elevation increases above sea level.

Automatic Sprinkler System — System of water pipes, discharge nozzles, and control valves designed to activate during fires by automatically discharging enough water to control or extinguish a fire. Also called sprinkler system.

Auxiliary Fire Alarm System — System that connects the protected property with the fire department alarm communications center by a municipal master fire alarm box or over a dedicated telephone line.

B

Backflow Prevention Device — Required device installed on the water supplies to fire protection systems to prevent potential contamination of the domestic water supply by cross-connection to a nondomestic source.

Boiling Point — Temperature of a substance when the vapor pressure exceeds atmospheric pressure. At this temperature, the rate of evaporation exceeds the rate of condensation. At this point, more liquid is turning into gas than gas is turning back into a liquid.

C

Carbonaceous — Made of or containing carbon.

Central Station Alarm System — System that functions through a constantly attended location (central station) operated by an alarm company. Alarm signals from the protected property are received in the central station and are then retransmitted by trained personnel to the fire department alarm communications center.

Certificate of Occupancy — Issued by a building official after all required electrical, gas, mechanical, plumbing, and fire protection systems have been inspected for compliance with the technical codes and other applicable laws and ordinances.

Churn – also sometimes referred to as operating at shutoff.

Circuit Breaker — Device (basically an on/off switch) designed to allow a circuit to be opened or closed manually, and to automatically interrupt the flow of electricity in a circuit when it becomes overloaded.

Class A Fire — Fire involving ordinary combustibles such as wood, paper, cloth, and similar materials.

Class B Fire — Fire of flammable and combustible liquids and gases such as gasoline, kerosene, and propane.

Class C Fire — Fire involving energized electrical equipment.

Class D Fire — Fire of combustible metals such as magnesium, sodium, and titanium.

Class K Fire — Fires in cooking appliances that involve combustible cooking media (vegetable or animal oils and fats). Commonly occurring in commercial cooking facilities such as restaurants and institutional kitchens.

D

Dead-End Hydrant — Fire hydrant located on a dead-end main that receives water from only one direction.

Deluge Sprinkler System — Fire-suppression system that consists of piping and open sprinklers. A fire detection system is used to activate the water or foam control valve. When the system activates, the extinguishing agent expels from all sprinkler heads in the designated area.

Dry Chemical Agent — (1) Any one of a number of powdery extinguishing agents used to extinguish fires. The most common include sodium or potassium bicarbonate, monoammonium phosphate, or potassium chloride. (2) Extinguishing system that uses dry chemical powder as the primary extinguishing agent; often used to protect areas containing volatile flammable liquids.

Dry-Pipe Sprinkler System — Fire-suppression system that consists of closed sprinklers attached to a piping system that contains air under pressure. When a sprinkler activates, air is released that activates the water or foam control valve and fills the piping with extinguishing agent. Dry systems are often installed in areas subject to freezing.

E

Elevation — (1) Height of a point above sea level or some datum point. (2) Drawing or orthographic view of any of the vertical sides of a structure or vertical views of interior walls.

Extinguishing Agent — Any substance used for the purpose of controlling or extinguishing a fire.

F

FM Global (FM) — Fire research and testing laboratory that provides loss control information for the Factory Mutual System and anyone else who may find it useful.

FFFP — Abbreviation for Film Forming Fluoroprotein Foam.

Fire Alarm Control Unit — A system component that receives input from automatic and manual fire alarm devices and may provide power to detection devices or communication devices. The fire alarm control unit can be a local control unit or a master control unit.

Fire Department Connection (FDC) — Point at which the fire department can connect into a sprinkler or standpipe system to boost the water flow in the system. This connection consists of a clappered siamese with two or more 2½-inch (65 mm) intakes or one large-diameter (4-inch [100 mm] or larger) intake. Also called Fire Department Sprinkler Connection.

Fire Extinguisher — Portable fire fighting device designed to combat incipient fires.

Fire Gas Detector — Device used to detect gases produced by a fire within a confined space.

Fire Pump — (1) Water pump used in private fire protection to provide water supply to installed fire protection systems. (2) Water pump on a piece of fire apparatus. (3) Centrifugal or reciprocating pump that supplies seawater to all fire hose connections aboard a ship.

Flame Detector — Detection and alarm devices used in some fire detection systems (generally in high-hazard areas) that detect light/flames in the ultraviolet wave spectrum (UV detectors) or detect light in the infrared wave spectrum (IR detectors). Also called Light Detectors.

Flammable Liquid — Any liquid having a flash point below 100°F (37.8°C) and having a vapor pressure not exceeding 40 psi absolute (276 kPa).

Flow Pressure — Pressure created by the rate of flow or velocity of water coming from a discharge opening. Also called Plug Pressure.

Foam — Extinguishing agent formed by mixing a foam concentrate with water and aerating the solution for expansion; for use on Class A and Class B fires. Foam may be protein, synthetic, aqueous film forming, high expansion, or alcohol type. Also known as Finished Foam.

Foam Concentrate — (1) Raw chemical compound solution that is mixed with water and air to produce foam. (2) The raw foam liquid as it rests in its storage container prior to the introduction of water and air.

Foam Proportioner — Device that injects the correct amount of foam concentrate into the water stream to make the foam solution.

Foam Solution — Result of mixing the appropriate amount of foam concentrate with water. Foam solution exists between the proportioner and the nozzle or aerating device that adds air to create finished foam.

Force — (1) To break open, into, or through. (2) Simple measure of weight, usually expressed in pounds or kilograms.

Frangible Bulb — Small glass vial fitted into the discharge orifice of a fire sprinkler. The glass vial is partly filled with a liquid that expands as heat builds up. At a predetermined temperature, vapor pressure causes the glass bulb to break, causing water to flow.

Friction Loss — That part of the total pressure lost as water moves through a hose or piping system, caused by water turbulence and the roughness of interior surfaces of hose or pipe.

Fusible Link — (1) Connecting link device that fuses or melts when exposed to heat. Used in sprinklers, fire doors, dampers, and ventilators. (2) Two-piece link held together with a metal that melts or fuses at a specific temperature.

G

Grid System — Water supply system that utilizes lateral feeders for improved distribution.

H

Halon — Halogenated agent; extinguishes fire by inhibiting the chemical reaction between fuel and oxygen.

Hazen-Williams Formula — Empirical formula for calculating friction loss in water systems; fire protection industry standard. To comply with most nationally recognized standards, the Hazen-Williams formula must be used.

Head Pressure — Pressure exerted by a stationary column of water, directly proportional to the height of the column.

Heat Detector — Alarm-initiating device that is designed to be responsive to a predetermined rate of temperature increase or to a predetermined temperature level.

Heat Release Rate (HRR) — Total amount of heat produced or released to the atmosphere from the convective-lift fire phase of a fire per unit mass of fuel consumed per unit time.

House Line — Permanently fixed, private standpipe hoseline.

Hydrostatic Test — A testing method that uses water under pressure to check the integrity of pressure vessels.

I

Impeller — Vaned, circulating member of the centrifugal pump that transmits motion to the water.

Inspection — Formal examination of an occupancy and its associated uses or processes to determine its compliance with the fire and life safety codes and standards.

L

Latent Heat of Vaporization — Quantity of heat absorbed by a substance at the point at which it changes from a liquid to a vapor.

Law of Specific Heat — (1) Measure of the heat-absorbing quality of a substance as measured in btu's or kilojoules. (2) Relative quantity of heat required to raise the temperature of substances or the quantity of heat that must be removed to cool a substance.

Local Alarm Systems — (1) Alarm systems that alert and notify only occupants on the premises of the existence of a fire so that they can safely exit the building and call the fire department. If a response by a public safety agency (police or fire department) is required, an occupant hearing the alarm must notify the agency. (2) Combination of alarm components designed to detect a fire and transmit an alarm on the immediate premises. Also known as protected premises fire alarm system.

M

Maintenance — Keeping equipment or apparatus in a state of usefulness or readiness.

N

Normal Operating Pressure — That pressure found in a water distribution system during normal consumption demands.

O

Occupancy — (1) General fire service term for a building, structure, or residency. (2) Building code classification based on the use to which owners or tenants put buildings or portions of buildings. Regulated by the various building and fire codes. Also called Occupancy Classification.

OS&Y Valve — Outside stem and yoke valve; a type of control valve for a sprinkler system in which the position of the center screw indicates whether the valve is open or closed. Also known as outside screw and yoke valve.

P

Plenum — An open space or air duct above a drop ceiling that is part of the air distribution system.

Polar Solvent — (1) Flammable liquids that have an attraction for water, much like a positive magnetic pole attracts a negative pole; examples include alcohols, ketones, and lacquers. (2) A liquid having a molecule in which the positive and negative charges are permanently separated resulting in their ability to ionize in solution and create electrical conductivity. Water, alcohol, and sulfuric acid are examples of polar solvents.

Post Indicator Valve — A type of valve used to control underground water mains that provides a visual means for indicating "open" or "shut" position; found on the supply main of installed fire protection systems. The operating stem of the valve extends above ground through a "post" and a visual means is provided at the top of the post for indicating "open" or "shut."

Preaction Sprinkler System — Fire-suppression system that consists of closed sprinkler heads attached to a piping system that contains air under pressure and a secondary detection system; both must operate before the extinguishing agent is released into the system; similar to the dry-pipe sprinkler system.

Pressure — Force per unit area exerted by a liquid or gas measured in pounds per square inch (psi) or kilopascals (kPa).

Pressure, water pressure — Pressure is defined as force per unit area in a liquid or gas.

R

Rate-of-Rise Heat Detector — Temperature-sensitive device that sounds alarm when the temperature changes at a preset value, such as 12°F to -15°F per minute.

Relief Valve — Pressure control device designed to eliminate hazardous conditions resulting from excessive pressures by allowing this pressure to bypass to the intake side of the pump.

Residual Pressure — Pressure at the test hydrant while water is flowing. It represents the pressure remaining in the water supply system while the test water is flowing and is that part of the total pressure that is not used to overcome friction or gravity while forcing water through fire hose, pipe, fittings, and adapters.

Required Delivered Density (RDD) — the amount of water needed at the fuel surface to suppress a given fire.

Response Time Index (RTI) — Numerical value representing the speed and sensitivity with which a heat responsive fire protection device (like a fusible link) responds.

Retard Chamber — Chamber that that catches and slows the excess water that may be sent through the alarm valve of an automatic sprinkler system during momentary water pressure surges. This reduces the chance of a false alarm activation. The retard chamber is installed between the alarm check valve and alarm signaling equipment.

Riser — (1) Vertical part of a stair step. (2) Vertical water pipe used to carry water for fire protection systems above ground such as a standpipe riser or sprinkler riser. (3) Pipe leading from the fire main to the fire station (hydrants) on upper deck levels of a vessel.

S

Saponification — A phenomenon that occurs when mixtures of alkaline based chemicals and certain cooking oils come into contact resulting in the formation of a soapy film.

Schedule — Wall thickness of pipe.

Siamese — Hose appliance used to combine two or more hoselines into one. The siamese generally has female inlets and a male outlet and is commonly used to supply the hose leading to a ladder pipe.

Smoke Detector — Alarm-initiating device designed to actuate when visible or invisible products of combustion (other than fire gases) are present in the room or space where the unit is installed.

Specific Gravity — Weight of a substance compared to the weight of an equal volume of water at a given temperature. A specific gravity less than 1 indicates a substance lighter than water; a specific gravity greater than 1 indicates a substance heavier than water.

Specific Heat — The amount of heat required to raise the temperature of a specified quantity of a material and the amount of heat necessary to raise the temperature of an identical amount of water by the same number of degrees.

Sprig-up — A pipe that rises vertically and supplies a single sprinkler.

Sprinkler — Waterflow device in a sprinkler system. The sprinkler consists of a threaded nipple that connects to the water pipe, a discharge orifice, a heat-actuated plug that drops out when a certain temperature is reached, and a deflector that creates a stream pattern that is suitable for fire control. Formerly referred to as a sprinkler head.

Static – At rest or without motion.

Static Pressure — (1) Potential energy that is available to force water through pipes and fittings, fire hose, and adapters. (2) Pressure at a given point in a water system when no water is flowing.

U

Underwriters Laboratories Inc. — Independent fire research and testing laboratory with headquarters in Northbrook, Illinois that certifies equipment and materials. Equipment and materials are approved only for the specific use for which it is tested.

V

Vacuum — Space completely devoid of matter or pressure. In fire service terms, it is more commonly used to describe a pressure that is somewhat less than atmospheric pressure. A vacuum is needed to facilitate drafting of water from a static source.

Velocity — Speed; the rate of motion in a given direction. It is measured in units of length per unit time such as feet per second (meters per second) and miles per hour (kilometers per hour).

Venturi Principle — Physical law stating that when a fluid, such as water or air, is forced under pressure through a restricted orifice, there is an increase in the velocity of the fluid passing through the orifice and a corresponding decrease in the pressure exerted against the sides of the constriction. Because the surrounding fluid is under greater pressure (atmospheric), it is forced into the area of lower pressure. Also called Venturi Effect.

Viscosity — Liquid's thickness or ability to flow.

Waterflow Alarm — Alarm-initiating device actuated by the movement (flow) of water within a pipe or chamber; most common installation is in the main water supply pipe of a sprinkler system. Also called Waterflow Detectors and Sprinkler System Supervision Systems.

Water Hammer — Force created by the rapid deceleration of water causing a violent increase in pressure that can be powerful enough to rupture piping or damage fixtures. It generally results from closing a valve or nozzle too quickly.

Water Supply — Any source of water available for use in fire fighting operations.

Wall Post Indicator Valve (WPIV) — Similar to a post indicator valve (PIV) but mounted on the wall of the protected structure.

Index

A

Accelerators, 187–189
Acceptance testing
    dry-pipe sprinkler systems, 204
    function, 76–77
    procedures, 77–78
    wet-pipe sprinkler system, 200
Activation of extinguishers, 40–42
Actual delivered density (ADD) of sprinklers, 173
ADD (actual delivered density) of sprinklers, 173
AFFF. *See* Aqueous film forming foam
Agents. *See* Extinguishing agents
Air pressurizing water (APW) extinguishers, 23
Aircraft hangar standards (NFPA 409), 227
Air-sampling smoke detector, 73–74
Alarm panel, 52
Alarm-initiating devices. *See also* Signaling systems
    air-sampling smoke detector, 73–74
    annunciated alarm, 59–60
    automatic, 66–76, 79–80
    bimetallic detector, 68–69
    coded pull stations, 65
    combination detector, 76
    continuous line detector, 68, 69
    deluge sprinkler system, 208–209
    double-action pull station, 65–66
    fire gas detector, 75–76, 80
    fixed-temperature heat detector, 66–69
    flame detector, 74–75, 80
    frangible bulbs, 67–68
    fusible links, 67–68, 80
    inspecting and testing, 79–80
    ionization smoke detector, 72–73
    manual, 64–66, 79
    noncoded pull stations, 65
    photoelectric smoke detector, 71–72
    pneumatic rate-of-rise line detector, 70, 80
    pneumatic rate-or-rise spot detector, 70, 80
    preaction sprinkler system, 208–209
    purpose, 54
    rate compensated detector, 70–71
    rate-of-rise heat detector, 69–71
    requirements, 64–65
    single-action pull station, 65
    smoke detectors, 71–74, 76
    thermoelectric detector, 71
    waterflow alarm test, 200–201
Alkaline mixture extinguishing agents, 15–16
Altitude, defined, 99
Annunciated alarm, 59–60
Annunciator panel, 59
Antifreeze
    extinguishing agent, 15, 30
    in wet-pipe sprinkler systems, 184–185
Application of extinguishing agent, 42
APW (air pressurizing water) extinguisher, 23

Aqueous film forming foam (AFFF)
    Class A fires, 43–44
    Class B fires, 45
    flammable liquid fires, 28, 29
    foam nozzles, 236, 238
    foam/water sprinklers, 239
    in foam-water systems, 230
    hydrostatic test, 37
    overview, 16–17
    techniques of extinguisher use, 43–44, 45–46
Around-the-pump proportioner, 235
Atmospheric pressure, 96–97
Automatic alarms, 66–76, 79–80
Automatic door closing devices, 55
Automatic nozzles, 239
Automatic sprinkler systems. *See* Sprinkler system
Automotive fire apparatus, 31
Auxiliary fire alarm system, 61, 81

B

Backflow prevention
    private water supply systems, 115–116
    valves, 178, 179
    water flow restriction, 135
Balanced pressure proportioner, 233–235
Ball valves, 175
Ball-drip valves, 179
Battery power supply, 53–54, 130
Bells as notification devices, 55
Bimetallic detector, 68–69
Boiling point, 89
Branch line, 175
Branches for hydrants, 114
Building and fire code requiring sprinklers, 166
Burning characteristics, 7–8
Butterfly valve
    control valve, 175
    function, 112–113
    indicator, 176
    rotating disc, 177
    in suction pipes, 134
Bypass line, 138–139

C

Calcium chloride, 88
Can-type pump installation, 126
Carbon dioxide extinguishing agents
    Class C fires, 46
    dangers, 223–224
    in gaseous systems, 223–226
    hydrostatic test, 37
    NFPA 12 standards, 223
    overview, 16
    specific gravity, 92
    specific heat, 88
    system components, 225–226
    techniques of extinguisher use, 45

warning signs, 224
water vs., 89
Cartridge-operated extinguishers
 activation, 41
 hydrostatic test, 37
 overview, 23–24
 recharging, 36, 37
Central station system, 62–63, 81
Centrifugal fire pump, 123, 124–125, 139
Certificate of occupancy, 78
Check valve, 135, 177–178, 181–182
Churches, 27
Circuit breaker, 131
Circulation relief valve, 136
Class I
 commodities, 193–194
 high-rise building standpipe systems, 156
 service, 117
 standpipe system fire department connections, 153
 standpipe system function, 152
 standpipe system water supply, 155
Class II
 commodities, 193–194
 service, 117
 standpipe system function, 152
 standpipe system water supply, 155
Class III
 commodities, 193–194
 high-rise building standpipe systems, 156
 service, 117
 standpipe system fire department connections, 153
 standpipe system function, 152–153
 standpipe system water supply, 155
Class IV commodities, 193–194
Class A fire
 burning characteristics, 7
 classification of fire, 10
 extinguishers. See Class A fire extinguishers
 letter-symbol system, 9, 10
 pictorial system, 8–9
Class A fire extinguishers
 distribution, 27–28
 extinguishing agents, 15, 16, 19, 21–22
 rating tests, 12
 requirements, 8
 size and placement, 26–27
 techniques of use, 43–45
 types of extinguishers, 40
Class B fire
 burning characteristics, 7
 classification of fire, 10
 extinguishers. See Class B fire extinguishers
 letter-symbol system, 9, 10
 operator guides, 13
 pictorial system, 8–9
Class B fire extinguishers
 distribution, 28–29
 extinguishing agents, 16, 18, 21
 rating tests, 12, 13
 requirements, 8
 techniques of use, 45–46
 types of extinguishers, 40

Class C fire
 burning characteristics, 7
 classification of fire, 10
 extinguishers. See Class C fire extinguishers
 letter-symbol system, 9, 10
 pictorial system, 8–9
Class C fire extinguishers
 distribution, 29
 extinguishing agents, 16, 18, 19, 21
 rating tests, 12, 14
 requirements, 8
 techniques of use, 46
 types of extinguishers, 40
Class D fire
 burning characteristics, 7
 classification of fire, 10
 extinguishers. See Class D fire extinguishers
 letter-symbol system, 9, 10
Class D fire extinguishers
 distribution, 29
 dry powders, 22–23
 rating tests, 12, 14
 requirements, 8
 techniques of use, 46–47
 types of extinguishers, 40
Class E fire, 7. See also Class K fire
Class K fire
 burning characteristics, 7
 classification of fire, 10
 extinguishers. See Class K fire extinguishers
 letter-symbol system, 9, 10
 pictorial system, 8–9
Class K fire extinguishers
 distribution, 29
 extinguishing agents, 15–16, 21
 rating tests, 14–15
 requirements, 8
 techniques of use, 47
 types of extinguishers, 40
Clean agents, 18–19, 20, 223
Cloud chamber air sampling smoke detector, 73–74
Coded pull stations, 65
Coefficient of friction, 100
Coefficients for testing, 144
Combination detector, 76
Combustible liquids code (NFPA 30), 227
Combustible metals. See Class D fire
Communication systems, 63
Complex loop, 110
Compressed-gas cylinders, 38, 39
Computer-aided signals, 60
Concentric reducer, 135
Continuous line detector, 68, 69
Control valves, 175–177
Controllers for pumps
 diesel engines, 133–134
 electric motors, 131–133
 function, 131
Coupled water motor-pump proportioner, 236
Critical velocity, 103
Cross main, 175
Cylinder-operated extinguishers, 37

D

Darcy-Weisbach technique, 118
Deluge sprinkler system
    applicability, 56
    inspections, 208–209
    restoration, 210
    testing, 208–209
    valve operation, 189–191
Depth of flammable liquid fires, 28–29, 46
Detection systems. *See also* Signaling systems
    alarm control unit, 52, 80–81
    auxiliary services, 55–56, 81
    flame detector, 74–75, 80
    heat. *See* Heat detectors
    initiating devices, 54
    light detectors, 74–75
    power supply, 52–54
    system components, 51–56
    testing, 51, 52
Diesel engine drivers
    controllers, 133–134
    cooling system, 130
    engine power, 129–130
    engine requirements, 130
    fuel storage, 130–131
    gear drives, 130
    maintenance, 131
    overview, 129
    testing, 147
    weekly inspections, 140, 147
Differential principle, 186–187
Discharge of fire extinguishers
    duration, 11–12
    hydrostatic test, 12
    range, 12
    volume capability, 11
Discharge orifice, 173
Distribution of fire extinguishers
    Class A, 27–28
    Class B, 28–29
    Class C, 29
    Class D, 29
    Class K, 29
    elements, 25
    nature-of-the-hazard factor, 26
    size-of-the-extinguisher factor, 26–27
Door closing devices, 55
Double interlock system, 191
Double-action pull station, 65–66
Drain valves, 178
Drum drip valve, 203
Dry chemical extinguishers
    Class B fires, 45
    Class C fires, 46
    Class D fires, 22–23, 46–47
    discharge volume capability, 11
    extinguishing agents, 19, 21–23, 35
    hydrostatic test, 37
    maintenance, 35
    monoammonium phosphate, 21–22, 44
    overview, 19, 21
    potassium bicarbonate, 21
    pressurizing, 36
    recharging, 35
    sodium bicarbonate, 21, 35, 88
    systems, 222–223
    techniques of extinguisher use, 44
Dry-pipe sprinkler system
    inspections, 203–204
    installation, 185
    operational sequence, 189
    quick-opening devices, 187–189
    restoration, 210
    testing, 204–207
    valves, 185–187

E

Early-suppression fast-response (ESFR) sprinkler, 170–171, 172–173
Eccentric reducer, 135
Eccentric wheel, 60
Electric motors for pumps
    acceptance test, 143–146
    controllers, 131–133
    overview, 127–129
    weekly inspections, 140, 147
Electrical deluge system, 190
Electrical equipment fires. *See* Class C fire
Electronic equipment extinguishing systems, 219, 223
Electronic relay circuit, 60
Elevated tanks, 109, 116–117
Elevation, defined, 99
Elevation head, 97
Elevation head pressure, 99
Emergency start lever with latch, 132
Emergency voice/alarm communications system, 63, 81
Engineered wet chemical system, 220
ESFR (early-suppression fast-response) sprinkler, 170–171, 172–173
Essentials of Fire Fighting manual, 87
Exhausters, 187–189
Expellant cartridges, 38, 39
Extinguishing agents. *See also* Special extinguishing systems
    alkaline mixtures, 15–16
    antifreeze, 15, 30
    application, 42
    aqueous film forming foam, 16–17
    carbon dioxide. *See* Carbon dioxide extinguishing agents
    Class A, 15, 16, 19, 21–22
    Class B, 16, 18, 21
    Class C, 16, 18, 19, 21
    Class K, 15–16, 21
    clean agents, 18–19, 20, 223
    dry chemical agents, 19, 21–23, 35
    film forming fluoroprotein, 17–18, 37, 239
    halons, 18. *See also* Halons
    reusing expelled agents, 36
    substitutions, 11
    water, 15, 30
Extra-hazard occupancy, 27

F

Factory Mutual System (FM Global)
    component testing, 51
    diesel engines, 129
    fire pump testing, 123
    standard performance specifications, 139
FDCs. *See* Fire department connections
Feed main, 175
FFFP. *See* Film forming fluoroprotein
Film forming fluoroprotein (FFFP)
    foam/water sprinklers, 239
    hydrostatic test, 37
    overview, 17–18
Fire department connections (FDCs)
    caps, breakable, 154
    function, 154, 180
    placement, 178, 179
    in sprinkler systems, 179–180
    in standpipes, 153–154
    wet-pipe sprinkler system, 180
    zoned, 154
Fire extinguishers. *See* Portable fire extinguishers
Fire gas detector, 75–76, 80
Fire Protection Hydraulics and Water Supply Analysis
    manual, 87, 146
Fire pumps
    can-type installation, 126
    centrifugal, 123, 124–125, 139
    circulation relief valve, 136
    coefficients for testing, 144
    component arrangement, 138–139
    controllers, 131–134
    diesel engines. *See* Diesel engine drivers
    drivers, 127–131
    electric motors. *See* Electric motors for pumps
    function, 123
    gauges, 138
    horizontal split-case. *See* Horizontal split-case fire
        pumps
    impeller rotors, 123, 124, 128
    installation standards. *See* NFPA 20
    maintenance, 125
    multiple-stage, 125
    performance specifications, 139
    pipe and fittings, 134–136
    positive-displacement, 123
    pressure maintenance, 137–138
    relief valves, 136
    single-stage, 124–125
    steam turbines, 131
    test curves, 141, 142
    test equipment, 136–137
    testing, 123, 136–137, 139–146, App. C
    vertically mounted split-case, 124, 125, 126
    vertical-shaft turbine, 125–127, 128, 138
    zone, 157
Fireplug, 108
Fixed foam system, 227, 228
Fixed-temperature heat detector
    bimetallic detector, 68–69
    continuous line detector, 68, 69
    frangible bulbs, 67–68

    fusible links, 67–68, 80
    placement, 66–67
    temperature classification, 67
Flame detector
    infrared, 74
    testing, 80
    ultraviolet wave spectrum, 74–75
Flammable liquid fires. *See also* Class B fire
    with depth, 28–29, 46
    Flammable and Combustible Liquids Code (NFPA 30),
        227
    techniques of extinguisher use, 45–46
    without depth, 28
Flow pressure, 99
Flushing underground connections, 200, 204
FM. *See* Factory Mutual System
FM-200, 19
Foam extinguishing systems
    applicability, 226–227
    around-the-pump proportioner, 235
    balanced pressure proportioner, 233–235
    coupled water motor-pump proportioner, 236
    fixed foam system, 227
    foam proportioners, 231–236
    foam/water system, 229–230, 239
    function, 230
    high-expansion foam system, 229
    inspections, 241
    line eductors, 231–233
    NFPA standards, 227
    nozzles and sprinklers, 236–241
    premixed foam, 236
    pressure proportioning tank system, 235
    production of foam, 230–231
    semifixed Type A system, 227–228, 229
    semifixed Type B system, 228
    testing, 241
Force, 92–94
FPP, *Fire Protection Hydraulics and Water Supply Analysis*
    manual, 87, 146
Frangible bulbs
    function, 67–68
    in sprinkler systems, 168, 169, 170
Freezing
    heat absorption of ice, 90
    problems during firefighting, 93
    property of water, 87, 92
    wet-pipe sprinkler system, 184–185, 200
Friction loss
    causes, 99–100, 101–103
    principles, 100–101
    reduction, 103
    water supply pipe, 108
Fusible links
    function, 67–68
    inspections, 80
    in sprinkler systems, 168, 169
    wet chemical extinguishing systems, 221

G

Gaseous extinguishing systems, 223–226
Gate valves, 112, 175–177

Gauges on fire pumps, 138
Gear drives, diesel engine, 130
Generator power supply, 54
Globe valves, 178–179
Gravity water system, 104–106, 155
Grid system for water distribution, 104, 107–108, 109, 110

H

Halons
    Class A fires, 44–45
    Class C fires, 46
    extinguishing agent, 18
    in gaseous systems, 223
    hydrostatic test, 37, 38
    pressurizing fire extinguishers, 36
    replacement agents, 18–19
Halotron, 19
HCFC (hydrochlorofluorocarbon), 19, 20
Head pressure
    described, 97
    elevation head pressure, 99
    table of head in feet, 98
Heat detectors
    bimetallic detector, 68–69
    combination, 76
    continuous line detector, 68, 69
    fixed-temperature, 66–69
    frangible bulbs, 67–68
    fusible links, 67–68, 80
    nonrestorable, 77, 80
    pneumatic rate-of-rise line, 70, 80
    pneumatic rate-of-rise spot, 70, 80
    rate compensated, 70–71
    rate-of-rise, 69–71, 80
    restorable, 77, 80
    thermoelectric, 71
HFC (hydrofluorocarbon), 19, 20
High back-pressure foam aspirator nozzles, 239, 240
High-expansion foam generator nozzles, 239–241
High-expansion foam system, 229
High-rise buildings, standpipes, 155–156, 157
Horizontal split-case fire pumps
    acceptance test, 143–147
    component arrangement, 138
    electric motors, 128
    illustrated, 123, 124, 125, 140
    overview, 123–125
Horns as notification devices, 55
Hose systems. *See also* Standpipe systems
    friction loss, 100–101
    function, 151
    water supplies, 117–118, 156
House line, 152
Hydrants
    bonnet, 113, 114
    branch, 114
    coefficients for testing, 144
    color codes, 115
    dead-end, 107
    dry barrel, 113–114
    function, 113
    looped, 107

maintenance, 114–115
    NFPA 291 standards, 115
    spacing, 114
    wet barrel, 113, 114, 115
Hydraulic deluge system, 190
Hydrochlorofluorocarbon (HCFC), 19, 20
Hydrofluorocarbon (HFC), 19, 20
Hydrostatic tests
    discharge pipe, 134
    dry-pipe sprinkler systems, 204
    fire pumps, 140
    intervals, 37
    portable fire extinguishers, 36–39
    rating criteria, 12
    record-keeping, 38–39
    standpipe systems, 159
    wet-pipe sprinkler system, 200
    wetting agent, 37

I

IFSTA
    Essentials of Fire Fighting manual, 87
    Pumping Apparatus Driver/Operator Handbook, 87
Impairment control of sprinkler systems, 210–211
Impeller rotor, 123, 124, 128
Indicating valves, 111
Inergen, 19
Infrared flame detector, 74
Inspections
    automatic alarm-initiating devices, 79–80
    carbon dioxide systems, 226
    deluge sprinkler system, 208–209
    dry-pipe sprinkler systems, 203–204
    fire detection and signaling system, 76–81
    fire extinguishers. *See* Inspections of portable fire extinguishers
    foam extinguishing systems, 241
    inspector's test valve, 178–179, 201
    manual alarm-initiating devices, 79
    preaction sprinkler system, 208–209
    records, 82–84
    sprinkler systems, 167
    standpipe systems, 159–160
    water-based systems (NFPA 25), 147–148
    wet chemical extinguishing systems, 221–222
Inspections of portable fire extinguishers
    impairments, 32
    maintenance, 32, 34–35
    procedures, 33–34
    record-keeping, 33
Insurance Services Office (ISO), 115
International Fire Code, hydrant spacing, 114
Inverting-type fire extinguishers, 25
Ionization smoke detector, 72–73
ISO (Insurance Services Office), 115
Isolation switch, 132

J

Jockey pump (pressure maintenance pump), 133

K

Kitchens. *See also* Class K fire
    extra-hazard occupancy, 27
    UL 300, *Standard for Fire Testing of Fire Extinguishing Systems for Protection of Restaurant Cooking Areas*, 14
    wet chemical extinguishing systems, 219–220

L

Labeling. *See* Signage on fire extinguishers
Latent heat of vaporization, 88, 89–90
Law of Latent Heat of Vaporization, 88, 89–90
Law of Specific Heat, 88–89
Leak test for fire extinguishers, 36
Learning objectives
    fire detection and signaling systems, 50
    fire extinguishing agents and portable fire extinguishers, 6
    fire pumps, 122
    special extinguishing systems, 218
    sprinkler systems, 163–164
    standpipes and hose systems, 150
    water supplies, 86
Letter-symbol system, 9, 10
Light detectors, 74–75
Light-hazard occupancy, 27
Line eductors, 231–233
Lith-X®, 23
Local energy system, 61
Local signaling systems, 57–59, 80, 81
Lock pins on fire extinguishers, 33
Loops, water distribution, 107, 109–110
Low-expansion foam nozzles, 237

M

Magnesium fire tests, 14
Main drain tests
    deluge sprinkler system, 208–209
    dry-pipe sprinkler systems, 204
    preaction sprinkler system, 208–209
    wet-pipe sprinkler system, 201–203
Maintenance
    diesel engines, 131
    fire detection and signaling system, 78
    fire extinguishers. *See* Maintenance of portable fire extinguishers
    hydrants, 114–115
    pump controllers, 131
    sprinkler systems, 167
    water-based systems (NFPA 25), 147–148
Maintenance of portable fire extinguishers
    defined, 32
    procedures, 34
    records, 38
    vehicle-mounted units, 35
Manual pull stations, 64–66
Master control unit, 57–59, 80–81
Metering device, 136, 137
Met-L-X®, 23
Mobile home sprinkler system standards. *See* NFPA 13D
Monoammonium phosphate dry chemical agent, 21–22, 44

Municipal water supply systems
    direct pumping system, 104–106
    distribution systems. *See* Water distribution systems
    gravity system, 104–106, 155
    inadequacy, 123
    means of moving water, 104–106
    sources, 104
    treatment facilities, 104, 106

N

NAFED. *See* National Association of Fire Equipment Distributors
Nameplate. *See* Signage on fire extinguishers
National Association of Fire Equipment Distributors (NAFED)
    fire extinguisher effectiveness study, 7
    pictorial system development, 8
    recharging properly, 35
National Electrical Code (NFPA 70), 51, 129
National Electrical Manufacturers Association (NEMA), 127
National Fire Alarm Code. See NFPA 72
Nature-of-the-hazard distribution factor, 26
NA-X®, 22–23
NEMA (National Electrical Manufacturers Association), 127
NFPA 10, *Standard for Portable Fire Extinguishers*
    depth of flammable liquid fire, 28
    extinguisher travel distance, 29
    inspections, 33
    letter-symbol system, 9
    pictorial system, 8–9
    reusing expelled agents, 36
    selection and operation, 7
NFPA 11, *Standard for Low- Medium- and High-Expansion Foam Systems*, 227
NFPA 12, *Standard on Carbon Dioxide Extinguishing Systems*, 223
NFPA 13, *Standard for the Installation of Sprinkler Systems*
    dry-pipe sprinkler systems, 185
    foam extinguishing systems, 227
    pipe schedule systems, 118
    pipe testing, 134
    preaction systems, 191
    provisions, 167
    quick-opening devices, 187
    restoring sprinkler systems, 210
    spare sprinkler requirements, 196
    storage areas, 193
    wet-pipe sprinkler system, 181
NFPA 13D, *Standard for the Installation of Sprinkler Systems in One- and Two-Family Dwellings and Mobile Homes*
    provisions, 167
    purpose, 211
    water supplies, 213, 214
    wet-pipe sprinkler system, 181
NFPA 13R, *Standard for the Installation of Sprinkler Systems in Residential Occupancies up to and Including Four Stories in Height*
    provisions, 167
    purpose, 211

water supplies, 213
wet-pipe sprinkler system, 181
NFPA 14, *Standard for the Installation of Standpipe and Hose Systems*
   automatic water supplies, 117
   classification of systems, 151–152
   hoselines, 156
   pressure-regulating devices, 158
   water pressure, 157
NFPA 16, *Standard for the Installation of Foam-Water Sprinkler and Foam-Water Spray Systems*, 227
NFPA 17, *Standard for Dry Chemical Extinguishing Systems*, 222
NFPA 17A, *Standard for Wet Chemical Extinguishing Systems*, 219, 221–222
NFPA 20, *Standard for the Installation of Stationary Pumps for Fire Protection*
   fire pump standards, 123
   metering devices, 136
   pump installation, 141
   relief valves, 136
   standard performance specifications, 139
   steam turbines, 131
   water flow restrictions, 135
NFPA 24, *Standard for the Installation of Private Fire Service Mains and Their Appurtenances*, 134
NFPA 25, *Standard for the Inspection, Testing, and Maintenance of Water-Based Fire Protection Systems*
   annual flow pump test, 147–148
   dry-pipe sprinkler systems, 185
   impairment control, 211
   main drain tests, 201, 202
   provisions, 167
   quick-opening devices, 187
   restoring sprinkler systems, 210
   vane-type water flow switch tests, 201
NFPA 30, *Flammable and Combustible Liquids Code*, 227
NFPA 70, *National Electrical Code*
   electric motors, 129
   provisions, 51
NFPA 72, *National Fire Alarm Code*
   central station system, 62
   heat-sensing fire detectors, 67
   inspections, 78
   power supply, 52–54
   provisions, 51
   pull stations, 64
   shunt system, 61
   system requirements, 57
NFPA 230, *Standard for the Fire Protection of Storage*, 167, 193
NFPA 231, *Standard for General Storage*, 118
NFPA 291, *Recommended Practice for Fire Flow Testing and Marking of Hydrants*, 115
NFPA 409, *Standard on Aircraft Hangars*, 227
NFPA 1901, *Standard on Automotive Fire Apparatus*, 31
NFPA 2001, *Standard on Clean Agent Fire Extinguishing Systems* (2004), 18–19
NFPA job performance requirements, App. A
Noncoded local alarm, 57, 58
Noncoded pull stations, 65
Nonindicating valves, 111–113

Non-interlock preaction sprinkler system, 191
Nonrestorable heat detectors, 77, 80
Normal operating pressure, 97, 99
Notification devices, 55, 183–184
Nozzles
   automatic, 239
   foam extinguishing systems, 236–241
   foam-water sprinkler, 239
   handline, 236
   high back-pressure foam aspirator, 239, 240
   high-expansion foam generator, 239–241
   low-expansion foam, 237
   self-educting foam, 238
   smoothbore, 237
   standard fixed-flow fog, 238–239
   water spray, 172

O
Obsolete extinguishers, 25
Occupancy
   certificate of, 78
   extra-hazard, 27
   light-hazard, 27
   ordinary-hazard, 27
   sprinkler inspections due to changes, 197
Occupational Safety and Health Administration (OSHA)
   fire extinguisher effectiveness study, 7
   hoseline training, 117
One-way voice alarm system, 63
Ordinary combustibles. *See* Class A fire
Ordinary-hazard occupancy, 27
OS&Y. *See* Outside stem and yoke (OS&Y) valve
OSHA. *See* Occupational Safety and Health Administration
Outside stem and yoke (OS&Y) valve
   function, 111
   illustrated, 112, 176
   for sprinkler systems, 175
   suction pipe, 134

P
Parallel telephone system, 61
Parmalee, Henry S., 165
P.A.S.S. operating method, 42–43
Pendant sprinklers, 171–172
Perfect vacuum, 97
Photoelectric smoke detector, 71–72
Pictorial system, 8–9
Pilot lamp, 132
Piping
   copper tubing, 173–174
   ferrous metal, 173
   galvanized, 173
   plastic, 173, 174
   playpipes, 142, 143
   polybutylene, 174
   residential sprinklers, 213
   schedule, 118, 173
   sprinkler inspections, 196–197
   sprinkler system, 173–175
   wet chemical extinguishing systems, 220–221
Pitot tube, 142, 143

PIV. *See* Post indicator valve
Plastics, 193–194
Playpipes, 142, 143
Pneumatic deluge system, 190
Pneumatic rate-of-rise line detector, 70, 80
Pneumatic rate-or-rise spot detector, 70, 80
Portable crash truck, 235
Portable fire extinguishers
    agents. *See* Extinguishing agents
    cartridge-operated. *See* Cartridge-operated
      extinguishers
    classification of fire on faceplate, 11
    discharge, 11–12
    effectiveness, 7
    on fire apparatus, 31
    hydrostatic tests, 36–39
    inspections, 32–34
    installation, 29–31
    maintenance, 32, 34–35
    obsolete, 25
    placement, 29–31
    pump-operated, 24
    ratings, 9–15, App. B
    recharging, 35–36
    selection and distribution, 24–29, 39–40
    standards. *See* NFPA 10
    stored-pressure. *See* Stored-pressure extinguishers
    techniques of use, 39–47
    UL 711 standards, 10, 14
    water mist, 46
    wheeled, 40, 42
Post indicator valve (PIV)
    function, 111, 112
    illustrated, 176
    for sprinkler systems, 175–176
    wet-pipe sprinkler system, 199
Potassium bicarbonate dry chemical agent, 21
Potassium fire tests, 14
Power supply to alarms
    batteries, 53–54, 130
    generators, 54
    local energy system, 61
    primary, 52
    secondary, 52–53, 54
    trouble signal, 53–54
Preaction sprinkler system
    applicability, 191
    double interlock system, 191
    inspections, 208–209
    non-interlock system, 191
    operational sequence, 191
    restoration, 210
    single interlock system, 191
    testing, 208–209
Pre-engineered wet chemical system, 220
Premixed foam, 236
Pressure
    altitude, 99
    atmospheric, 96–97
    balanced pressure proportioner, 233–235
    defined, 92
    elevation, 99

    flow, 99
    force vs., 92–94
    friction loss, 99–103
    head, 97, 98, 99
    maintenance pump, 133
    normal operating, 97, 99
    principles, 94–96
    pump tests, 143
    residual, 99
    static, 97–99
    switch, 132, 201
    velocity, 99
    velocity head, 94
    water hammer, 103, 202
    water pressure-regulating devices, 158–159
Pressure gauge on fire extinguishers, 34
Pressure maintenance pump, 137–138
Pressurization of fire extinguishers, 36, 40, 41
Primary power supply, 52
Priming water, 203, 204
Private water supply systems, 115–117, 134
Projected beam smoke detectors, 72
Proportioning foam
    around-the-pump proportioner, 235
    balanced pressure proportioner, 233–235
    coupled water motor-pump proportioner, 236
    line eductors, 231–233
    premixed foam, 236
    pressure proportioning tank system, 235
Proprietary system, 62, 81
psi, 97
psia, 97
psig, 97
Public water supply. *See* Municipal water supply systems
Pull boxes, 64–66
Pumping Apparatus Driver/Operator Handbook, 87
Pumping stations, 109
Pump-operated extinguishers, 24
Pumps. *See* Fire pumps
Purple K, 21

Q
Quick-opening devices, 187–189
Quick-response sprinkler, 170–171

R
Rate compensated detector, 70–71
Rate-of-rise heat detector, 69–71
    pneumatic rate-of-rise line detector, 70, 80
    pneumatic rate-of-rise spot detector, 70, 80
    rate compensated detector, 70–71
    thermoelectric detector, 71
Rating portable fire extinguishers
    Class A rating tests, 12
    Class B rating tests, 12, 13
    Class C rating tests, 12, 14
    Class D rating tests, 12, 14
    Class K rating tests, 14–15
    criteria, 11–12
    discharge duration, 11–12
    discharge range, 12
    discharge volume capability, 11

hydrostatic test, 12
overview, 9–11
rating tests, App. B
UL 711, *Standard For Rating and Fire Testing Fire Extinguishers*, 10, 14
RDD (required delivered density) of sprinklers, 173
Recharging fire extinguishers, 35–36
Recommended Practice for Fire Flow Testing and Marking of Hydrants (NFPA 291), 115
Records
    electronic, 83–84
    fire detector tests, 80
    fire extinguisher inspections, 33
    hydrostatic tests, 38–39
    inspections, 82–84
    public domain, 82
    storage length, 82
    written, 82–83
Reducers, 135
Refractory photoelectric smoke detector, 72
Relief valve, 136
Remote receiving system, 61–62, 81
Required delivered density (RDD) of sprinklers, 173
Residential sprinkler system
    application barriers, 212
    conventional sprinklers vs., 212–213
    design layout, 214
    design modifications, 212
    fatality statistics, 211
    NFPA standards. *See* NFPA 13D; NFPA 13R
    piping, 213
    spacing, 214
    water supplies, 213–214
Residual pressure, 99
Restorable heat detectors, 77, 80
Risers, 151, 156, 174
Running period timer, 132

S

Safety issues
    carbon dioxide, 16
    diesel engines, 130
    electric motors, 132, 144
    extinguishing agent substitutions, 11
    flammable liquid fires, 28
    foam extinguishers, 17
    halons, 18
    playpipes, 143
    pump controller maintenance, 131
    rotating equipment, 145
    standpipe system for fire attack, 152
    water extinguishers, 46
    water-based extinguishers, 43
Saponification, 16, 219
Sears Tower standpipe system, 157
Secondary power supply, 52–53, 54
Selection of fire extinguishers
    elements, 25, 39–40
    nature-of-the-hazard factor, 26
    size-of-the-extinguisher factor, 26–27
Selective coded system, 60
Self-educting foam nozzles, 238

Semifixed Type A System, 227–228, 229
Semifixed Type B system, 228
Shunt system, 61
Siamese connections, 179. *See also* Fire department connections
Sidewall sprinklers, 172
Signage on fire extinguishers
    factory test pressure, 38
    fire classes, 11
    inspections, 34
Signaling systems. *See also* Alarm-initiating devices; Detection systems; Smoke detectors
    activation, 39, 40
    auxiliary fire alarm system, 61, 81
    auxiliary services, 55–56, 81
    central station system, 62–63, 81
    control unit, 52, 53, 80–81
    emergency voice/alarm communication, 63, 81
    factors in choice of, 56
    function, 56
    local energy system, 61
    local system, 57–59, 80, 81
    master control unit, 57–59, 80–81
    noncoded local alarm, 57, 58
    notification appliances, 55, 183–184
    parallel telephone system, 61
    proprietary system, 62, 81
    remote receiving system, 61–62, 81
    selection factors, 56
    shunt system, 61
    zoned/annunciated alarm, 59–60
Simple loops, 109–110
SIN (Sprinkler Identification Number), 210
Single interlock preaction sprinkler system, 191
Single-action pull stations, 65
Size-of-the-extinguisher factor, 26–27
Smoke detectors
    air-sampling smoke detector, 73–74
    cloud chamber air sampling, 73–74
    combination, 76
    dangers of smoke, 71
    ionization smoke detector, 72–73
    limitations, 74
    operation, 71
    photoelectric smoke detector, 71–72
    projected beam, 72
    refractory photoelectric smoke detector, 72
    tube-type air sampling smoke detector, 74
    visible products-of-combustion, 71–72
Smoothbore nozzles, 237
Smothering action, 91, 230
Snap action disk thermostat, 69
Soda-acid fire extinguishers, 25
Sodium bicarbonate dry chemical agent
    overview, 21
    recharging extinguishers, 35
    specific heat, 88
    uses for, 21
Sodium fire tests, 14
Speakers as notification devices, 55

Special extinguishing systems
    dry chemical, 222–223. *See also* Dry chemical
       extinguishers
    foam. *See* Foam extinguishing systems
    gaseous, 223–226
    purpose, 219
    wet chemical, 219–222. *See also* Wet chemical
       extinguishing systems
Specific gravity, 92
Specific heat, 88–89
Split-case pumps, vertically mounted, 124, 125, 126. *See also* Horizontal split-case fire pumps
Sprig-up, 174
Sprinkler Identification Number (SIN), 210
Sprinkler system
    actual delivered density (ADD), 173
    components, 166–173
    deflectors, 171–172, 173
    deluge. *See* Deluge sprinkler system
    design considerations, App. E
    discharge orifice, 173
    dry-pipe system. *See* Dry-pipe sprinkler system
    duration of water supplies, 118–119
    early-suppression fast-response (ESFR), 170–171, 172–173
    fire department operations, App. D
    flushed or concealed, 171
    heat-sensitive device, 168
    high-rise building standpipes, 155–156
    history, 165
    hydraulic, 118
    impairment control, 210–211
    inspections, 194–209
    occupancy changes, 197
    old-style, 171
    pendant, 171–172
    pipe schedule systems, 118
    piping, 173–175
    preaction system, 191
    protective cage, 171
    quick-response, 170–171
    reasons for use, 165–166
    reliability, 165
    required delivered density (RDD), 173
    residential. *See* Residential sprinkler system
    response time, 169–171
    restoring, 210–211
    sidewall, 172
    sprinkler head, 168
    sprinkler inspections, 195–196
    standards. *See* NFPA 13; NFPA 13D; NFPA 13R
    temperature ratings, 168–169, 170
    testing, 194–209
    thermal sensitivity, 173
    upright, 171, 172
    use of fire department connections, 179–180
    valves, 175–179
    water spray nozzles, 172
    water supplies, 117, 118–119, 156–157
    wet system, 167
    wet-pipe system. *See* Wet-pipe sprinkler system
Stairwell protection, 156

Standard fixed-flow fog nozzles, 238–239
Standard for Dry Chemical Extinguishing Systems (NFPA 17), 222
Standard for Fire Testing of Fire Extinguishing Systems for Protection of Restaurant Cooking Areas (UL 300), 14
Standard for General Storage (NFPA 231), 118
Standard for Low- Medium- and High-Expansion Foam Systems (NFPA 11), 227
Standard for Portable Fire Extinguishers. See NFPA 10
Standard For Rating and Fire Testing Fire Extinguishers (UL 711), 10, 14
Standard for the Fire Protection of Storage (NFPA 230), 167, 193
Standard for the Inspection, Testing and Maintenance of Water-Based Fire Protection Systems. See NFPA 25
Standard for the Installation of Foam-Water Sprinkler and Foam-Water Spray Systems (NFPA 16), 227
Standard for the Installation of Private Fire Service Mains and Their Appurtenances (NFPA 24), 134
Standard for the Installation of Sprinkler Systems. See NFPA 13
Standard for the Installation of Sprinkler Systems in One- and Two-Family Dwellings and Mobile Homes. See NFPA 13D
Standard for the Installation of Sprinkler Systems in Residential Occupancies up to and Including Four Stories hi Height. See NFPA 13R
Standard for the Installation of Standpipe and Hose Systems. See NFPA 14
Standard for the Installation of Stationary Pumps for Fire Protection. See NFPA 20
Standard for Wet Chemical Extinguishing Systems (NFPA 17A), 219, 221–222
Standard on Aircraft Hangars (NFPA 409), 227
Standard on Automotive Fire Apparatus (NFPA 1901), 31
Standard on Carbon Dioxide Extinguishing Systems (NFPA 12), 223
Standard on Clean Agent Fire Extinguishing Systems (2004), (NFPA 2001), 18–19
Standpipe systems
    automatic dry system, 153
    automatic wet system, 153
    fire department connections, 153–154
    fire department operations, App. D
    function, 151
    in high-rise buildings, 155–156, 157
    horizontal, 151
    inspections, 159–160
    manual dry system, 153
    manual wet system, 153
    primed system, 153
    semiautomatic dry system, 153
    testing, 159–160
    types, 153
    vertical, 151
    water pressure, 157–159
    water pressure-regulating devices, 158–159
    water supplies, 117–118, 155–157
Start and stop electric motor buttons, 132
Static pressure, 97–99
Steam turbines, 131
Storage areas

bulk paper, 192, 226–227
commodity classifications, 193–194
extra-hazard occupancy for paint, 27
high-piled materials, 192
NFPA 230 standards, 167, 193
NFPA 231 standards, 118
sprinkler systems, 167, 192–194
storage methods, 194
warehouse, App. F
Stored-pressure extinguishers
activation, 40, 41
hydrostatic test, 37, 38
leak tests, 36
overview, 23
Storz-type couplings, 154
Strobe lights as notification devices, 55
Strobe-type tachometer, 145
Study of fire extinguisher effectiveness, 7
Supervising station alarm system, 61
Symbols
Class A fire, 7. *See also* Class A fire
Class B fire, 7. *See also* Class B fire
Class C fire, 7. *See also* Class C fire
Class D fire, 7. *See also* Class D fire
Class K fire, 7. *See also* Class K fire
System riser, 151, 156, 174

T
Telephone signaling system, 61, 63
Testing
acceptance. *See* Acceptance testing
automatic alarm-initiating devices, 79–80
carbon dioxide systems, 226
deluge sprinkler system, 208–209
diesel engines, 140, 147
discharge pipes, 134
dry-pipe sprinkler systems, 204–207
electric motors, 140, 143–146
fire detection and signaling system, 51, 52, 76–81
fire pumps, 123, 136–137, 139–146, App. C
foam extinguishing systems, 241
hydrostatic test of portable fire extinguishers, 36–39
main drain tests. *See* Main drain tests
manual alarm-initiating devices, 79
potassium fire tests, 14
preaction sprinkler system, 208–209
shop test curves, 141, 142
sodium fire tests, 14
sprinkler systems, 167
standpipe systems, 159–160
system timetables, 81
trip test. *See* Trip test
UL 711, *Standard For Rating and Fire Testing Fire Extinguishers*, 10, 14
water-based systems (NFPA 25), 147–148
waterflow alarm test, 200–201
wet chemical extinguishing systems, 221–222
wet-pipe sprinkler system, 198–203
Thermoelectric detector, 71

Trip test
deluge sprinkler system, 208–209
dry-pipe sprinkler systems, 204–207
preaction sprinkler system, 208–209
Trouble signal power supply, 53–54
Tube-type air sampling smoke detector, 74

U
UL. *See* Underwriters Laboratories Inc.
UL 300, *Standard for Fire Testing of Fire Extinguishing Systems for Protection of Restaurant Cooking Areas*, 14
UL 711, *Standard For Rating and Fire Testing Fire Extinguishers*
potassium fire tests, 14
rating system, 10
sodium fire tests, 14
Ultraviolet wave spectrum flame detector, 74–75
Underwriters Laboratories Inc. (UL)
component testing, 51
diesel engines, 129
distilled water fire extinguisher, 15
fire pump testing, 123
Halotron rating, 19
listed fire pumps, 125
playpipes, 142
residential sprinklers, 212
UL 300, *Standard for Fire Testing of Fire Extinguishing Systems for Protection of Restaurant Cooking Areas*, 14
UL 711, *Standard For Rating and Fire Testing Fire Extinguishers*, 10, 14
United States Fire Administration (USFA), 211
Upright sprinklers, 171, 172
USFA (United States Fire Administration), 211

V
Vacuum, 97
Valves
alarm check valve, 181–182
automatic drain, 178
backflow prevention, 178, 179
ball, 175
ball-drip, 179
butterfly. *See* Butterfly valve
check valve, 135, 177–178, 181–182
circulation relief, 136
control, 175–177
deluge, 189–191
drain, 178
drum drip, 203
dry-pipe, 185–187
function, 111
gate, 112, 175–177
globe, 178–179
indicating, 111
inspector's test valve, 178–179, 201
keys, 111
nonindicating, 111–113
operating, 177–179
OS&Y. *See* Outside stem and yoke (OS&Y) valve
PIV. *See* Post indicator valve
preaction, 191
relief, 136

underground, 111
valve clapper pivot, 178
wall post indicator valve (WPIV), 175–176
waterflow switch, 182–183
Vane-type waterflow switch, 183, 201
Vehicles
NFPA 1901, *Standard on Automotive Fire Apparatus*, 31
vehicle-mounted extinguishers, 31, 35
Velocity head pressure, 94
Velocity pressure, 99
Venturi principle, 231, 232
Vertical-shaft turbine pumps
applicability, 125–127
component arrangement, 139
gauges, 138
illustrated, 141
impeller assembly, 128
Viscosity of liquids, 91
Visible products-of-combustion smoke detector, 71–72
Voice/alarm communications system, 63, 81

W

Wall post indicator valve (WPIV), 175–176
Warehouse storage, App. F. *See also* Storage areas
Water
for automatic sprinkler systems, 118–119
carbon dioxide vs., 89
characteristics, 87–92
columning, 187
distribution. *See* Water distribution systems
duration of supplies, 118–119
as extinguisher. *See* Water extinguishing agent
gravity system, 104–106, 155
hammer, 103, 202
for hose systems, 117–118, 156
ice, 87, 90, 92, 93
municipal. *See* Municipal water supply systems
pressure, 92–99
private systems, 115–117, 134
properties, 87
residential sprinklers, 213–214
smothering action, 91
specific heat, 88–89
for sprinkler systems, 156–157
standpipe system pressure, 157–159
for standpipe systems, 117–118, 155–157
steam, 87, 90–91
supply main, 174
surface area, 90–91
temperature and state of, 87, 88
waterflow alarm test, 200–201
Water distribution systems
backflow prevention. *See* Backflow prevention
complex loops, 110
dead-end hydrants, 107
distributors, 107–108
elevated storage tanks, 109, 116–117
grid system, 104, 107–108, 109, 110
hydrants, 113–115

looped hydrants, 107
pipe diameter, 108–109
piping arrangement, 109
piping materials, 108
primary feeders, 107–108
private, 115–117, 134
pumping stations, 109
secondary feeders, 107–108
simple loops, 109–110
valves, 111–113
Water extinguishing agent
advantages of use, 92
Class C fires, 46
disadvantages of use, 92
Law of Latent Heat of Vaporization, 88, 89–90
Law of Specific Heat, 88–89
physical environment, 30
properties, 15, 87–92
specific gravity, 92
standards. *See* NFPA 25
surface area, 90–91
techniques of use, 43–45
Water mist extinguisher, 46
Water motor gongs, 201
Water spray nozzles, 172
Waterflow alarm test, 200–201
Waterflow switch, 182–183
Wet chemical extinguishing systems
applicability, 219
hydrostatic test, 37
inspections, 221–222
NFPA 17A standards, 219, 221–222
pre-engineered, 220
system design, 220–221
testing, 221–222
Wet-pipe sprinkler system
alarm check valve, 181–182
antifreeze, 184–185
fire department connections, 180
freezing damage, 200
inspections, 198–203
installation, 181
main drain test, 201–203
notification systems, 183–184
restoration, 210
retard chamber, 182
testing, 198–203
waterflow alarm test, 200–201
waterflow switch, 182–183
Wetting agent hydrostatic test, 37
Wheeled fire extinguishers, 40, 42
Wire-type continuous line heat detector, 68
WPIV (wall post indicator valve), 175–176

Z

Zones
alarms, 59–60
fire department connections, 154
fire pumps, 157